Oiling the Wheels
of Apartheid

Oiling the Wheels
of Apartheid
Exposing
South Africa's
Secret Oil Trade

Arthur Jay Klinghoffer

Lynne Rienner Publishers • Boulder & London

Published in the United States of America in 1989 by
Lynne Rienner Publishers, Inc.
1800 30th Street, Boulder, Colorado 80301

and in the United Kingdom by
Lynne Rienner Publishers, Inc.
3 Henrietta Street, Covent Garden, London WC2E 8LU

Library of Congress Cataloging-in-Publication Data
Klinghoffer, Arthur Jay, 1941-
    Oiling the wheels of apartheid: exposing South Africa's secret
oil trade / Arthur Jay Klinghoffer
        p. cm.
    Bibliography: p.
    Includes index.
    ISBN 1-55587-164-X (alk. paper)
    1. Petroleum industry and trade—Political aspects—South Africa.
2. Economic sanctions—South Africa. 3. Apartheid—South Africa.
I. Title.
HD9577.S632K55   1989
382'.42282'0968—dc19   89-30731
                                                        CIP

British Cataloguing in Publication Data
A Cataloguing in Publication record for this book
is available from the British Library.

Printed and bound in the United States of America

The paper used in this publication meets the requirements
of the American National Standard for Permanence of
Paper for Printed Library Materials Z39.48-1984.

To Libby Jones Klinghoffer
and the memory of Rose Jones

with love and gratitude toward my matrilineage

CONTENTS

___ ACKNOWLEDGMENTS ___

Research for this book was carried out during a year's leave from Rutgers University under the auspices of the Faculty Academic Study Program. Significant financial support was also given by the university's President's Coordinating Council on International Programs, and the London School of Economics provided a comfortable and intellectually stimulating research environment. Special thanks must be extended to the Shipping Research Bureau (Amsterdam), the Holland Committee on Southern Africa (Amsterdam), and the Anti-Apartheid Movement (London), all of which provided valuable assistance. Michael Siegel is to be commended for his fine maps.

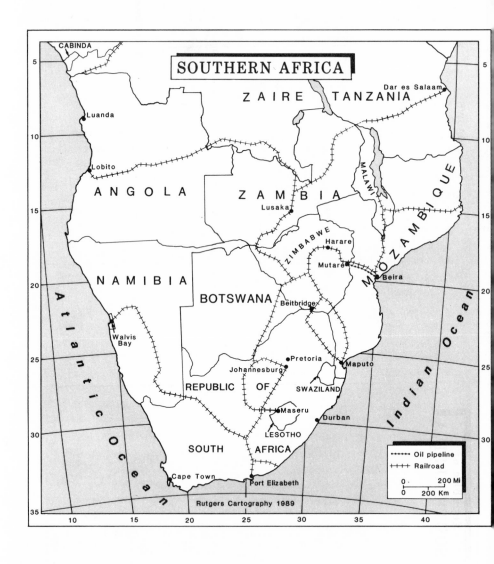

SOUTHERN AFRICA

CABINDA

ZAIRE   TANZANIA

Dar es Salaam

Luanda

Lobito

ANGOLA   ZAMBIA

Lusaka

MALAWI

MOZAMBIQUE

NAMIBIA   ZIMBABWE   Harare

BOTSWANA   Mutare
Beitbridge   Beira

Walvis
Bay

Pretoria
Johannesburg   Maputo
REPUBLIC   OF   SWAZILAND
Maseru   Durban
LESOTHO

SOUTH   AFRICA

Cape Town   Oil pipeline
Port Elizabeth   Railroad

0   200 Mi
0   200 Km

Rutgers Cartography 1989

Atlantic Ocean

Indian Ocean

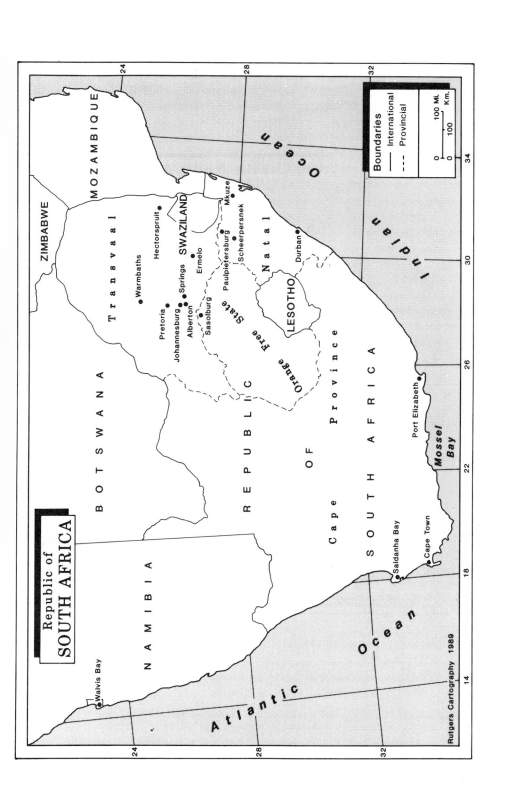

Republic of
SOUTH AFRICA

ZIMBABWE

MOZAMBIQUE

Transvaal

Warmbaths

Hectorspruit

SWAZILAND

Mkuze

Pretoria

Johannesburg · Springs
Alberton
Sasolburg

Ermelo

Paulpietersburg

Scheerpersnek

Natal

Orange Free State

Durban

LESOTHO

REPUBLIC

OF

SOUTH AFRICA

Cape   Province

Port Elizabeth

Mossel Bay

BOTSWANA

NAMIBIA

Walvis Bay

Saldanha Bay

Cape Town

Indian Ocean

Atlantic Ocean

Boundaries
——— International
- - - - Provincial

100 Ml.

100 Km.

0

0

Rutgers Cartography 1989

14   18   22   26   30   34

24   28   32

24   28   32

1

HIGH COMBUSTION
Oil and the
Antiapartheid Strategy

In the wee morning hours of Monday, June 2, 1980, as most South Africans lay asleep, antiapartheid guerrillas cut through security fences and set off explosive charges alongside three Natref oil refinery storage tanks in Sasolburg, forty miles south of Johannesburg, and in five tanks nearby, owned by Sasol, a corporation engaged in converting coal to oil. Firemen responded quickly but the total damage was estimated at almost $9 million: eight tanks had become inoperative and a considerable quantity of oil had been lost. The guerrillas avoided detection and disappeared into the night. At the same time, other saboteurs set explosives at a Sasol plant in Secunda, more than 150 miles away. Only a small fire resulted, and damage was minimal. In Springs, just outside Johannesburg, two bombs were placed at a recruiting office of Fluor, a U.S. company assisting with the construction of Sasol plants; the bombs were defused by security personnel.

That same day, the African National Congress claimed responsibility for these strikes against South Africa, and ANC president Oliver Tambo indicated that he had known of the planned sabotage in advance.[1] Not coincidentally, these attacks on South African oil facilities were made just after the International Day for an Oil Embargo Against South Africa, which was endorsed by the ANC and observed on May 20.

The struggle against apartheid entered a new stage as oil installations came to be considered prime targets. Though the South African government moved decisively to bolster security, more attacks followed. In June 1981 a limpet mine was discovered at a Shell depot in Alberton. That October an explosion rocked a

1

new Sasol plant under construction at Secunda. In 1982 attacks were made on a pipeline at Scheepersnek and fuel depots at Paulpietersburg, Hectorspruit, Mkuze, and Durban; storage facilities at Warmbaths were hit in October 1983. The following year explosives were detonated at Ermelo; five tanks were extensively damaged and the ANC headquarters in Lusaka, Zambia, claimed responsibility. Two months later four guerrillas were killed by the police after firing a rocket at the Mobil refinery in Durban. In November 1985 rockets were again used in an assault on two Sasol plants at Secunda, and in June 1986 leakage and fires resulted from the sabotage of a pipeline that connected the Mobil and Sapref refineries near Durban.[2]

Opponents of apartheid see oil as an Achilles heel of the South African system because there are no commercially viable petroleum deposits there—some oil is secured through the coal-to-oil conversion process, but most has to be imported. The African National Congress sees oil as synonymous with South African military power because security forces are highly mechanized—they use close to 30,000 barrels of oil per day.[3] Oil facilitates domestic oppression, the occupation of Namibia, and until recently, forays into Angola; therefore, actions obstructing its supply are seen as blows directed at one of the most crucial tools of apartheid.

Sabotage is a highly dramatic—but largely symbolic—tactic in the black liberation struggle. Also significant is the anti-Shell campaign, which developed over more than a decade, but has been most efficiently mobilized and carefully orchestrated since January 1986. (Shell is targeted because of its important refining and marketing role within South Africa and because it is accused of furnishing oil products for the military.)[4] But the most basic component of the oil offensive is the effort to impose and sustain an embargo. As expressed by the ANC: "The ability of the South African regime to maintain its aggressive actions against neighbors, and to use its repressive apparatus against the oppressed majority in South Africa is and will continue to be related directly to the quantity of oil the regime is able to obtain from outside South Africa."[5]

As the battle over apartheid has taken on new dimensions, the oil embargo has been adopted as a weapon in the antiracist arsenal. Imposed in conjunction with a moral crusade, the embargo has not succeeded in restricting oil flow to South Africa, however, because those supposedly applying the embargo are not complying with it. Though South African countermeasures surely play some role, the general failure to implement oil sanctions effectively results basically from the hypocrisy of much of the international community. Governments, oil companies, and shipping lines routinely take the moral high road in public, condemning apartheid and even endorsing the oil embargo and claiming adherence to its principles. Clandestinely, however, they take the low road of pecuniary interest. Considerable moral scrutiny has rightfully been directed at South Africa's racial policies, and more than the first stone has already been cast in that direction. It is now time to turn the ethical mirror around to examine those who throw such stones.

2

APPLYING PRESSURE
Implementing Oil Sanctions

Many states and international organizations have gone beyond verbal assaults on apartheid and have proclaimed oil embargoes against South Africa. Steps taken include bans on the delivery of crude oil products and petroleum technology and prohibitions on the provision of finance capital. Personnel and know-how may also be restricted, including assistance with the development of alternative energy sources or conversion processes (such as Sasol's production of oil from coal). Measures restricting tankers plying the South African oil trade and corporations that facilitate Pretoria's purchases of crude or play a role in refining and marketing are also possible.

To evaluate the effectiveness of the oil weapon, it is imperative to ascertain: Who is officially maintaining an embargo, and who isn't? How are the aims of the oil embargo related to broader sanctions directed at South Africa?

ORGANIZATIONAL MOMENTUM

Calls for an oil embargo against South Africa were issued at conferences of African leaders at least as early as 1960. In May 1963 the Organization of African Unity (OAU) was established as the continent's diplomatic forum for independent states, and the July 1964 summit endorsed general oil sanctions directed at apartheid. In November 1973, during the course of another Arab-Israeli war, African efforts against South Africa became entwined with Middle Eastern policies. Arab states sought black African backing against

5

Israel, while the Africans saw an opportunity to secure Arab support for an oil embargo against South Africa and possibly some discount purchases of Arab oil. The trade-off that emerged found all black African states except Malawi, Swaziland, and Lesotho breaking diplomatic relations with Israel. At an OAU Council of Ministers session held November 19-21 in Addis Ababa, Ethiopia, a "Resolution on Co-operation Between African and Arab States" called for Arab (and Iranian) oil embargoes against South Africa, Rhodesia, and Portugal. (Rhodesia was included as a target because it was then ruled by a white minority government whose proclamation of independence was not recognized by Britain or most other states; Portugal still maintained colonial control over the territories of Angola, Mozambique, and Guinea-Bissau.) Another resolution advocated an oil embargo of Israel, and there was discussion about Afro–Arab financial consultation on oil agreements and the creation of an Arab bank to assist black African states with their oil payments. Arab League members within the OAU were supportive, and OAU secretary-general Nzo Ekangaki was instrumental in working out this arrangement.[1]

The Algiers Arab summit of November 26-28 followed the OAU's lead: A decision was made to break all political, consular, economic, and cultural ties with South Africa, and an oil embargo was imposed on South Africa, Rhodesia, and Portugal; a separate resolution enacted an embargo against states supporting Israel. Futhermore, as expressed in a resolution on Africa, "the heads of state decided to convey greetings and appreciation to the fraternal African states for their decisions to break off relations with Israel."[2] This diplomatic deal was most effective in its early stages, but black African resentment later grew as it became apparent that Arab states were unwilling to offer oil at preferential prices or to provide the degree of financial assistance anticipated.

The Arab embargo against Israel's allies ended in March 1974, but the embargo of South Africa was never terminated. Implemented by the Organization of Arab Petroleum Exporting Countries (OAPEC), a group of Arab suppliers founded in January 1968, the embargo was in fact strengthened in Kuwait with a May 6, 1981 resolution calling for the blacklisting of tankers that had visited South African ports and the denial of oil to companies delivering there. Either or both penalties could be applied dependent upon "the size and character of the violation."[3]

The OAU stepped up the pressure on South Africa in 1977 by establishing a Committee of Seven to gain the cooperation of the Organization of Petroleum Exporting Countries (OPEC), founded in September 1960. Major Arab producers such as Saudi Arabia, Kuwait, and Iraq were members of both OPEC and OAPEC, but six of OPEC's thirteen members were non-Arab: Some supported neither the 1973-1974 embargo against pro-Israeli states nor the one directed at South Africa; non-Arab Iran was in fact the main supplier of both Israel and South Africa. The Committee of Seven included all four African OPEC members (Nigeria, Gabon, Algeria, Libya) as well as Ghana, Sierra Leone, and Gambia. Headed by Zambian foreign Minister Siteke Mwale, the committee sent delegations to all OPEC countries in 1977, with the exception of Iran, which refused to serve as host. All other OPEC members individually endorsed the embargo, agreed to adhere to it diligently, and continued to vote for it at the United Nations. (OPEC is purely an economic organization and does not officially endorse an embargo against South Africa, but individual members are free to arrive at their own judgments.)

Since 1978, the OAU has advocated very specific mechanisms for enforcing the embargo, and although African states have played a major role at the United Nations in calling for such measures, some states have at times sold oil directly to South Africa. African states have also agreed not to sell oil to companies dealing with South Africa, but this policy too is often disregarded. As for OPEC, its members have verbally backed oil sanctions since the Iranian revolution, and the Arab embargo is still on the books, yet South Africa receives most of its oil from Arab states and Iran. In addition, the policy of OPEC members that tankers calling at South Africa are banned from their ports is honored more in the breach than in the observance.[4]

Whereas Communist-ruled countries do not have to confront the embargo issue—they have not ever furnished oil to South Africa—Western states have enacted embargoes full of loopholes. Norway and Denmark, rich in North Sea crude, banned sales, and the foreign ministers of the entire European Economic Community (EEC) voted in September 1985 to follow suit, effective the end of January 1986. Nevertheless, implementing this policy was no simple matter: Denmark prohibited its ships from delivering oil, but it was unclear how jurisdiction could be applied once ships were chartered to noncitizens. Rotterdam, the busiest oil port in Europe, applied sanctions in November 1985, but could not really keep track of the transshipment of crude taking place through its storage tanks. Britain continued to provide oil products and did little to control the operations of its oil companies or tankers, although the Commonwealth has been imposing sanctions since

October 1985.

In practice, European embargoes have served as constraints rather than barriers—the oil continues to flow. Most of it is non-European in origin, but European shipping and oil companies participate in its delivery, claiming that to do otherwise would prove financially disastrous and result in the loss of thousands of jobs. On the other side of the Atlantic, U.S. crude and tankers play a very minor role in the South African oil trade, although U.S. oil companies are certainly active in refining and marketing within South Africa. In October 1986 the United States banned the provision of both crude and oil products and even made it illegal for Americans to sell foreign oil to South Africa; the regulations did not apply to contracts previously negotiated, however.

Not only international organizations and governments have taken steps against apartheid; workers have also reacted. In Trinidad and Tobago, members of the Oilfields Workers' Trade Union began to expose their country's links to South Africa as early as 1977, revealing that ships transporting oil products to South Africa were docking and acquiring fuel; the workers voiced suspicions that oil was being loaded on tankers as well. Their efforts led to the appointment of a commission of inquiry in January 1979, and an investigation of Texaco's oil operations ensued.[5] In March 1980 Libya set the stage for protest by hosting the International Oilworkers' Anti-Monopoly Conference, which called for the application of oil sanctions against South Africa. The following February Libyan port workers refused to load the Liberian-flag combined carrier *Atlantic Courage* because it had transported South African coal the previous year. The ship had been chartered by a U.S. company, Coastal States, which had no connection to the coal incident, but the firm lost $1 million because it had to charter another tanker to pick up its Libyan cargo.[6] In 1982 the union representing 9,000 British workers of the Shell International Trading Company (Sitco) called upon the company to halt deliveries to South Africa, charging that even though Shell was no longer supplying Omani oil directly, it was facilitating its provision through intermediaries. Sitco countered that it was only part of the Shell group and was not responsible for the actions of affiliates such as Shell South Africa.[7]

A sanctions conference sponsored by the International Labor Organization in June 1983 suggested workers take anti-South African actions; British, Australian, and Danish seamen's unions took the lead in forming Maritime Unions Against Apartheid (MUAA) in February 1984 and encouraging workers to support the oil embargo. At a Conference of Maritime Trade Unions in October 1985, a MUAA statement proclaimed: "The maritime unions can take action to increase the risks to shippers of oil and to increase the cost of oil to South Africa. Coordinated action on boycotts and delayed sailings can provide an important contribution to the enforcement of the oil embargo."[8] African National Congress president Oliver Tambo addressed the conference and endorsed these tactics; Jim Slater, general secretary of Britain's National Union of Seamen, pointed out that causing shipping delays of just one or two days could be effective. Companies would gain a reputation for unreliability, chartering out tankers would become more difficult, and extra days in port would increase costs.[9] That December the Seamen's Union of Australia boycotted two vessels because their owners had furnished tankers to deliver oil products from the United States to South Africa.[10]

The oil embargo has been publicized through numerous forums, including a March 1980 conference in Amsterdam organized by the Holland Committee on Southern Africa and a Dutch Christian antiapartheid movement known as Working Group Kairos. The UN Special Committee Against Apartheid also helped with the planning, and its chairman, B. Akporode Clark of Nigeria, attended. Messages of support were received from UN Secretary-General Kurt Waldheim and General Assembly President Salim Ahmed Salim. Specific steps were advocated for implementing the embargo, such as monitoring tanker movements so that ships supplying South Africa could be seized at their next port of call; boycotting by dock workers of offending vessels; and punishing by governments of companies that violated sanctions policies.[11] Antiapartheid organizations, the United Nations, and the OAU thereafter collaborated in pressing the embargo with important conferences held in Paris in May 1981, Amsterdam in September 1985, and Oslo in June 1986. Sympathetic West European parliamentarians gathered in Brussels in January 1981 and The Hague in November 1982.

Campaigns against Shell and Barclays Bank were organized. Shell had provided crude for South Africa and was instrumental in that country's refining and marketing operations. Barclays Bank controlled the largest banking network in South Africa and, in November 1979, it purchased shares in the development of Sasol oil-from-coal facilities. Barclays also assisted in the sale of other shares.[12] The campaign against Shell has been marginally successful—Shell no longer supplies South Africa with oil. The campaign against Barclays has been more effective: the bank announced in November 1986 a plan to divest itself of South African assets.

Extremely significant in the oil embargo effort is the role of the Shipping Research Bureau (Shirebu). Established in 1980 in Amsterdam by the Holland Committee on Southern Africa and Working Group Kairos, it includes two members from each organization on its four-person executive committee. Shirebu, directed by Jaap Woldendorp, maintains a close working relationship with the UN Special Committee Against Apartheid and serves as the hub of the system for monitoring South Africa's oil trade. By carefully tracking tanker movements and sifting information provided by crewmen and port workers, Shirebu has prepared numerous detailed studies of South African oil imports. These reports must be ranked as highly accurate because Shirebu painstakingly distinguishes between actual and suspected deliveries as it tracks the tankers, shipping companies, charterers, oil companies, and producing states involved. Shirebu also furnishes a wealth of information about the methods used to circumvent the embargo and the status of sanctions legislation in various countries.

DIPLOMATIC DISSENSION

The United Nations—ever the victim of world political divisions, caution, and even hypocrisy on the part of some members—has been ineffective in implementing an oil embargo of South Africa. A November 1963 General Assembly resolution on Namibia included oil sanctions against South Africa, but the issue lay relatively dormant until December 1977, when a specific resolution on South Africa was passed that included bans on deliveries and investments in the oil sector. The Special Committee Against Apartheid also commissioned Martin Bailey and Bernard Rivers to prepare a detailed report on Pretoria's oil situation. Their "Oil Sanctions Against South Africa," published in June 1978, focused increased attention on the issue.[13] In September the Special Committee Against Apartheid called upon the Security Council to impose a mandatory embargo, as did the General Assembly in January 1979, by a vote of 105 to 6 (with 16 abstentions). The negative votes were cast by the United States, Britain, France, West Germany, Belgium, and Luxembourg; Canada joined the opposition that December on another embargo resolution, and the seven negative votes remained the same in the December votes of both 1980 and 1981.

The General Assembly's resolutions grew increasingly detailed—for example, the December 1980 proposal attempted to cut off financial assistance to Sasol plants, to ban South Africa from acquiring shares in external oil development projects, and even to authorize the seizure of tankers that had delivered oil to South Africa—and were overwhelmingly favored by members. But the General Assembly, powerless to act under the UN Charter, could assert only a moral influence, and its pleas to the Security Council were never translated into mandatory sanctions.

In December 1982 the General Assembly asked the Special Committee Against Apartheid to appoint a committee of experts to report on oil sanctions, but its composition did not bode well for a strong embargo policy. Members were chosen from Algeria, Indonesia, Kuwait, Libya, Mexico, Nigeria, Norway, Romania, Saudi Arabia, and Venezuela—many of these countries were frequent violators of the embargo. December 1983 brought another General Assembly resolution (opposed by the usual six countries), but the next two years the embargo issue was covered under a more comprehensive legislative rubric rather than in separate resolutions.

In November 1986 Belgium and Luxembourg broke with the antiembargo faction, but Israel joined it in a 126 to 5 (with 15 abstentions) vote. An Intergovernmental Group to Monitor the Supply and Shipping of Oil and Petroleum Products to South Africa was established, chaired by Tom Eric Vraalsen of Norway and composed of representatives of eleven states, including OPEC members Algeria, Kuwait, and Nigeria. (At the time, Norwegian tankers were actively breaking the embargo, so the selection of Vraalsen was somewhat ironic.) When this committee's report was issued in November 1987, it did not include any independent research and instead relied exclusively on studies already published by the Shipping Research Bureau. The committee was cautious: Though it recommended Security Council consideration of a mandatory embargo, there was an inherent reluctance to blame states supplying South Africa or large oil companies, and, criticism was instead directed primarily at small companies and middlemen.[14]

Although the General Assembly has consistently supported an oil embargo, many states voting yes have indicated they will not actually apply sanctions unless the Security Council imposes a mandatory embargo—an unlikely step because the United States, Britain, and France have used their veto power since November 1977 to block Security Council action. It is not that these countries dispute the Security Council's right to mandate sanctions; after all, a precedent for an oil embargo against South Africa was established when one was directed at Rhodesia in 1966. The UN Charter is somewhat vague about the implementation of sanctions, but they appear to be an acceptable tactic as long as the target state is deemed a "threat to the peace".[15] Underlying the actions of the three Western powers may possibly be the fear of countersanctions by South Africa, or the possibility that similar tactics would later be directed at Israel. In recent years, there has also been a reluctance to squeeze South Africa too tightly because what the United States calls "constructive engagement" has become the cornerstone of antiapartheid policies. Thus noncompulsory sanctions rather than mandatory embargo have been recommended. In July 1985 the Security Council voted 13-0-2 (France joined the majority while the United States and Britain abstained) to urge states to suspend new investment, bar the marketing of krugerrands, restrict sports and cultural relations, prohibit new contracts in the nuclear field, and ban the sale of computer equipment that could be used by the South African military and police.[16]

UN rhetoric is often belied by actions. Apartheid is roundly condemned, but steps to counter it are taken circumspectly. Some states drag their feet; others, including many Arab producers and Iran, feint forward motion as they secretly seek profits by supplying oil. Norway, which endorsed the embargo, began to abstain from General Assembly votes once it became a major producer; Singapore, a transshipment point in the South Africa trade, was absent at some critical times of decision. Self-interest prevails as tankers proceed toward Durban and Cape Town against the background din of diplomatic diatribes and posturing.

TRADE AND POWER

Sanctions could have a major impact on South Africa. Its economy is well developed and closely interconnected with world markets, which permits other states to exert considerable influence. But sanctions can always turn into a two-edged sword—the more that economic ties are reduced, the less the potential to exert similar trade pressure in the future. In addition, target states tend to counteract sanctions by turning toward autarky, thus strengthening their ability to withstand external pressure. Consequently, economic sanctions are a difficult path to political ends because they arouse "a sense of community and solidarity" in the face of outside interference in a state's affairs.[17]

Economic sanctions are usually applied when diplomacy proves insufficient to the task. They may take the form of import, export, or financial sanctions, though only the last two tactics are specifically germane to the oil embargo. Financial sanctions bear no direct relationship to military force, but export controls could possibly be supplemented by armed might through the imposition of a naval blockade, a "hard sanction." In the South African case, the emphasis is on "soft sanctions" based on economic pressure and diplomacy.[18] History demonstrates, however, that sanctions are generally ineffective when not accompanied by force because full "economic warfare" must be applied. But the Security Council has not gone so far. It is true that the Security Council did unanimously vote for an arms embargo against South Africa in November 1977, and advocates of the oil embargo maintain that the denial of fuel is likewise necessary to weaken Pretoria's military potential. Nevertheless, the oil continues to flow, and no steps have even been initiated in regard to imposing a naval blockade.

Activating economic sanctions can be highly problematic, especially if an effort is made to apply universal rather than unilateral or multilateral sanctions. The universal approach has been used rarely (Italy in 1935 and Rhodesia in 1966 were unusual cases) because securing a consensus among the participants on goals and methods is difficult and because the profit motive often causes some of the sanctioning states to break ranks. Once this happens, others tend to follow suit if it is in their pecuniary self-interest. Ironically, the more states involved, the less the likelihood for success.[19] Even if a universal mandate for sanctions against South Africa could be obtained in the form of a Security Council vote, it would not assure greater effectiveness—some Western states in the strongest position to enforce sanctions through the use of a blockade are not strongly committed to the need for sanctions in the first place.[20] Consequently, it is important to realize that the number of participants who apply sanctions is less significant than the "degree of trade withdrawn from the receiver."[21]

Oil sanctions against South Africa have lacked major impact because they are not total. There is no blockade, companies supplying oil or tankers are rarely penalized, and there are no accompanying sanctions targeting South Africa in other areas such as finance, investment, and trade. An additional problem is that sanctions applied incrementally permit the targeted state to react, and South Africa's modernized economy has demonstrated its ability to do so effectively.[22] The country has managed to secure a sufficient supply of oil, and its white minority community has so far been unwilling to enact political reforms in response to outside economic pressure. In the long run, stronger sanctions may be due to increasing international pressure and South Africa, a strongly anti-communist state, will not have a practical option (as Cuba did in 1960-61) to shift diplomatically to the Soviet camp in an effort to buttress its economy. But South Africans can use its gold and strategic minerals to threaten countersanctions against the industrialized states, and it can try to shift its economic burden onto neighboring states that are highly dependent on South Africa's transportation and financial infrastructure. The big question mark will be the attitude of black South Africans, whose general support for sanctions under present economic conditions will likely weaken if the economy is forced to constrict.

THE SANCTIONS WEAPON

Sanctions are a response to "a breach of legal obligation."[23] In this context, the UN Security Council has determined that they are applicable in the South African case (at least in the form of an arms embargo plus selected economic measures) because Pretoria represents an ostensible threat to international peace and security. This interpretation is conditioned by the UN Charter, which disallows sanctions imposed on the basis of moral disapproval of a state's internal system. Sanctions against South Africa therefore have the legal imprimatur of the United Nations, although the oil embargo specifically does not.

Sanctions have objectives that parallel those of criminal law: to punish, to rehabilitate, to deter.[24] Punishment is a reprisal for a violation of international norms. If it is the only aim in this situation, South African blacks will bear much of the burden. Rehabilitation implies an effort to force changes in the South African system, resulting in compliance with international norms. However, the perceived transition from world outlawry to citizenship is vaguely conceptualized: Does it require only the elimination of apartheid, does it also include the removal of President Pieter Botha and other key officials, or does it depend upon a total dismantling of the entire South African political structure?[25] If control passes to black politicians, is this step sufficient for the termination of sanctions—or must power be held specifically by the African National Congress? In regard to deterrence, sanctions may be viewed as a warning that any extension or intensification of apartheid will not be tolerated by the world community.

Sanctions may be implemented positively, through promises and rewards, or negatively, through threats and punishments. However, it is often rather difficult to distinguish one approach from the other. Rewards may be proffered gradually, but then withdrawn for lack of compliance. The withholding of a reward thus becomes a form of punishment, just as a threat to punish a target state implies an eventual reward when compliance results in cessation of punishment.[26] The South African sanctions are primarily negative; sticks take precedence over carrots. Johan Galtung's general observation on sanctions made more than two decades ago appears highly germane: "If compliance is not obtained, there is at least the gratification that derives from knowing (or believing) that the sinner gets his due, that the criminal has been punished. In this sense negative sanctions are safer than positive sanctions. And when hatred is strong, positive sanctions would probably be out of the question anyway."[27]

Sanctions have not rehabilitated South Africa, the possibility of deterrence is still conjectural, and punishment has so far been minimal. The application of sanctions has been weak and ineffective—not surprising, considering they have not even been accompanied in most cases by the breaking of diplomatic relations. In practice, sanctions are mainly symbolic.

Yet symbolic sanctions demonstrate to the citizens of the sanctioning state that morality is being upheld and action taken; the domestic audience is as much a target as is the state being sanctioned. The U.S. boycott of the 1980 Moscow Olympics had no impact on Soviet policy in Afghanistan, but it did symbolize to U.S. citizens that a crusade was being waged against communist intervention. In fact, the implementation of largely symbolic sanctions is an effective way to disguise a government's inability to take concrete measures regarding the primary objective, the outlaw state. Secondary objectives include the effects of sanctions on the imposing state; tertiary objectives deal with the impact on the international community.[28] As Galtung wryly comments: "When military action is impossible for one reason or another, and when doing nothing is seen as tantamount to complicity, then something has to be done to express morality, something that at least serves as a clear signal to everyone that what the receiving nation has done is disapproved of. If the sanctions do not serve instrumental purposes, they can at least have expressive functions."[29]

Sanctions are a cautious and slow method of countering perceived illegality because force will tend to be the prime tactic when the practitioner seeks rapid and definitive results. Fredrik Hoffmann has perceptively written: "When sanctions are used, the manifest goal will probably not be attained, because the very decision to apply sanctions probably indicates that the motivation of the sanctioning country is relatively low."[30] On the effectiveness scale, symbolic sanctions rank lowest, followed by those precipitating economic inconvenience, severe economic deprivation, and total compliance.[31] Sanctions directed against South Africa have barely reached the second stage, namely economic inconvenience, though operating costs have indeed increased. But many organizations attempting to apply these sanctions have limited means at their disposal and cannot act as forcefully as can states, such as the United States, Britain, or France. Organizations like Working Group Kairos or the Holland Committee on Southern Africa may already be stretched to the limit of their capabilities.

3

THE ROOTS OF VULNERABILITY
South Africa's Energy Sector

Oil is critical to several sectors of the South African economy, and with the exception of the crude produced from coal via the Sasol process, Pretoria is completely dependent on imports. Legislation mandating secrecy, in an effort to conceal procurement methods from public scrutiny, has also been effective in cloaking the refining and marketing operations of major Western oil companies that have been crucial to South Africa's energy system. Clearly, the embargo has been easily circumvented and thus has had little impact on the oil supply.

POWERING APARTHEID

South Africa's energy-intensive economy is about 75 percent dependent on coal; inexpensive domestic production is possible because of the low wages paid to black miners. Nuclear energy has increased since the start-up of the Koeberg reactor in March 1984, but it still represents only a negligible percentage of total energy usage. Oil accounts for approximately 20 percent, and roughly one-third of Pretoria's outlays for imports go toward its acquisition. The crude oil deposits that have been located are not commercially viable. The only oil produced within South Africa, therefore, comes from the Sasol coal-to-oil technology, which can now provide about one-third of the country's needs (including the provision of bunkers to shops in transit). The original Sasol conversion plant is located in Sasolburg in the Orange Free State; Sasol II and III are at Secunda in the eastern Transvaal.

Although oil is used for only one-fifth of total energy consumption, it is extremely important in selected areas. The military and police are reliant on fuel produced from crude, and aviation fuel in particular is so essential that South Africa is probably paying the highest price in the world to purchase it. Little diesel fuel can be extracted from Sasol because its process is geared overwhelmingly toward the production of gasoline,[1] which makes railroads and other key economic sectors dependent on the refining of imported crude. Transportation is approximately 75 percent reliant on oil products, and autos obviously run on gasoline. Oil is also indispensable in the chemical, plastic, and fertilizer industries and in the production of commercial lubricants.

South Africa prohibits the release of statistics pertaining to its oil sector, but imports may be estimated at 325,000 barrels per day (310,000 b/d crude; 15,000 b/d oil products); Sasol's production adds another 95,000 barrels per day. Of the total, about 70,000 b/d are generally stockpiled; 40,000 b/d are exported to Namibia and neighboring black states; 50,000 b/d are supplied as bunkers for ships (including those in South Africa's own merchant marine); and 10,000 b/d are either lost in the refining process or used to power energy plants. This leaves 250,000 barrels per day for consumption.[2] South Africa imports 85 percent of its oil at Durban, where tankers under 100,000 deadweight tons may dock; larger vessels may use the offshore buoy that is connected to the mainland by pipeline. The remainder of the imports is handled at Cape Town and Saldanha Bay.

CONTRIBUTIONS AND CONFIDENTIALITY

Western companies serve as the backbone of South Africa's oil system. They provide the exploration and drilling for Pretoria's continuing search for resources, as well as the financing and technology needed to serve the oil sector. British and West German banks have been particularly active, as have U.S., French, and West German engineering firms.[3] The distribution of oil is dominated by five major oil companies, which control 85 percent of the market. British Petroleum (BP) and Shell play the largest roles, with the South African subsidiary of Shell being institutionally linked to the British rather than the Dutch branch of that company. Also important are Mobil and Caltex (both U.S.) and Total (French). Shell, Mobil, and Caltex are private corporations; BP and Total are partially owned by their respective states. Britain has gradually been privatizing BP, a process accelerated during Margaret Thatcher's administration, and the state share of ownership has now been reduced to less than one-third. France has a 40 percent stake in Total, which in turn controls almost two-thirds of Total South Africa. It should be added that BP and Total also have coal investments in South Africa, as does Shell.

The external role is crucial to the refining industry because sufficient capacity has been produced not only to meet needs but also to furnish bunkers and to export to neighboring states. Though the oil companies operating the refineries no longer supply their own crude they do process shipments arriving in South Africa and so circumvent the embargo; the result is indirect support of apartheid. Shell and BP jointly own the single buoy mooring off Durban, and they operate South Africa's largest refinery, Sapref (212,000 b/d). Also in Durban is a Mobil refinery (100,000 b/d), and a pipeline connects the port to the Natref refinery (75,000 b/d) in Sasolburg. Thirty percent of Natref is owned by Total, with 52.5 percent belonging to Sasol and 17.5 percent to the National Iranian Oil Company.[4] Caltex has a refinery (100,000 b/d) in Cape Town. A U.S. company, Fluor, has been instrumental in supplying technology for the Caltex and Natref refineries, as well as for the Sasol II plant in Secunda.

In order to cope with embargo efforts, South Africa seeks to hide its susceptibility to pressure and to protect the roles of those contributing to sanctions busting. Toward these ends, oil import and export statistics have not been published since late 1973 (originally in reaction to the Arab embargo), and tanker calls at South Africa ports are no longer listed. The keynote has been oil secrecy, the rudimentary guidelines of which first appeared in the National Supplies Procurement Act of 1970. They became much more comprehensive in 1977 with the passage of the Petroleum Products Act, which was strengthened in 1979 and again in 1985. Basically, it is illegal to publish information regarding the source, refining, transport, storage, stock level, or destination of oil; violators are subject to punishment of seven years in jail plus the possibility of a 7,000 rand fine. (When the law was instituted, the rand was more than a dollar, but is worth considerably less now.) Enforcement reached as far as Jaap Marais, leader of the Herstigte National Party, who disclosed in March 1981 that South Africa was selling gasoline to Zimbabwe. He was found guilty, but sentencing was waived.[5] Oil companies operating in South Africa are forced to comply with the secrecy legislation if they want to continue doing business there. Their parent corporations overseas therefore have limited knowledge of their South African subsidiaries, which provides a basis for denying responsibility for their actions.

Despite its campaign to maintain secrecy, South Africa has experienced considerable leakage of information. Some may even be gleaned from South African sources, such as the transcript of parliamentary debates and the publication Financial Mail.[6] Studies released by the Shipping Research Bureau and the UN Centre Against Apartheid shed great light on the subject, as do reports issued by the African National Congress and various antiapartheid organizations. Calculations about shipping movements are facilitated by reference to the publications of Lloyd's. Based upon the available evidence, it is apparent that South Africa's main asset is also its greatest potential liability: supportive efforts by numerous states and companies. Because South Africa's oil sector is extremely dependent on external assistance, a more effective embargo could lead to considerable economic dislocation.

4

COUNTERING THE CHALLENGE
Coping with the Oil Embargo

Sanctions have been imposed so gradually and over such a long time that South Africa has had ample opportunity to plan and implement countermeasures. Despite its failure to locate commercially exploitable domestic oil deposits, the Pretoria government has coped successfully with the oil embargo and has taken major steps to reduce reliance on an unstable external supply. Although claims about Sasol's output and the quantity of oil stored in strategic reserves tend to be exaggerated, South Africa has dealt with its problem in a multidimensional fashion, and it has recently been abetted by low crude oil prices engendered by the world oil glut. Nevertheless, oil sanctions have significantly raised South Africa's energy costs, making the burden to be borne not deprivation but excessive expenditure.

THE QUEST FOR OIL

South Africa has been unable to find any significant amounts of oil. The Southern Oil Exploration Corporation (Soekor), partially owned by Sasol, was founded in 1965 to seek out new deposits; it had little success—ninety of the ninety-one holes it drilled over its first eight years came up dry.[1] The search was accelerated after the Arab embargo of 1973-1974, but more natural gas than oil was discovered, and even the gas was not found in commercially exploitable quantities. Compounding difficulties was the political problem of Namibia's uncertain future, which prevented the development of the Kudu gas field off the Namibian coast. Oil exploration was undertaken both offshore and on land, but little effort has been expended on land since 1978. Foreign companies, as well as some private South African concerns, joined in the search, but they have generally pulled out as failures have mounted.[2] Soekor alone has spent close to 700 million rands in its quest; its yearly outlay in the mid-1980s was about 60 million rands.[3]

South Africa was so desperate for oil that it fell victim to the "sniffer plane" fraud, a purported technological breakthrough enabling an aircraft to locate offshore deposits. The Pretoria administration claims that no government funds were lost because the investor was Zululand Oil Exploration, a private company, which contracted with a subsidiary of the French firm Elf-Aquitane. Officially, 4.9 million rands were wasted, but the figure could actually be at least four times that amount.[4]

"Sniffer Plane" investments were made during the late 1970s, but they remained secret until revealed by a French publication in December 1983. The inventor of the electronic exploration equipment was an Italian, Aldo Bonassoli, and Elf-Aquitane invested in the project in June 1976. It bought the process in the fall of 1978, and president Giscard d'Estaing approved it as a clandestine operation of the French government. He even attended a demonstration of the "sniffer plane" in April 1979, but by July an adviser to the Ministry of Industry had determined the process was a fraud, and Elf-Aquitane withdrew after suffering a loss of 65 million pounds.[5] Though South Africans played only a minor role in the "sniffer plane" affair, it was indicative of their tendency to embrace anything that might promise oil.

One important potential source of oil is Mossel Bay, about 250 miles east of Cape Town. Discovery of a major gas deposit there was announced in January 1981, and plans were developed to convert some of this gas to 25,000 barrels per day of gasoline and diesel fuel. In November 1985 President Botha declared that the project would go ahead; it is estimated that production could begin in 1991. However, there are two serious obstacles: (1) Continued low oil prices would prevent the gas-to-oil process from being competitive economically; and (2), South Africa lacks the necessary technology and would have to turn to outside sources, which sanctions could potentially obstruct. Despite these important considerations, South Africa is likely to proceed for strategic reasons, taking its chances that the technology will be provided by British and U.S. firms. The British have particular expertise based on their experiences in North Sea development, and there are indications that the Thatcher government is encouraging private British support for the Mossel Bay project.[6]

Although South Africa has had little success at home, it is quietly building up significant energy assets overseas. The largest South African conglomerate, the Anglo American Corporation, has taken the lead by working through partially owned subsidiaries to acquire exploration rights in Canada, Indonesia, Britain, and the United States. In Britain, the Conservative government (in the face of Labour Party opposition) tried to ease access to North Sea crude. The plan was to permit offshore rights for its takeover of shares in Selection Trust, a company owned in part by the Anglo American-controlled Charter Consolidated. The deal, however, fell through.[7] No oil from any of these foreign tracts is actually exported to South Africa, but antiapartheid organizations are troubled by this overseas expansion into the oil sector and fear that it provides an eventual channel for circumventing oil sanctions.

South Africa's prime means of coping with an oil deficiency has been Sasol's oil-from-coal process. Sasol, the South African Oil and Gas Corporation, was originally a government-financed parastatal, but it later sold shares publicly (some purchased by British banks since 1979) and is now only partially state-owned. Organized in 1947, Sasol soon decided to convert coal to oil; the first plant (in Sasolburg) began operations in 1955 with a capacity of 5,000 barrels per day. As aptly described by Willie Breytenbach: "Today, when people reflect back on the 1950 decision to build the Sasol oil-from-coal plant, at the time considered an economic white elephant, they attribute it to foresight. It could, equally, have been a manifestation of that laager mentality. Economic development in the last 30 years has been based on the bedrock of strategic self-sufficiency, even at the cost of economic viability."[8]

In December 1974 plans were made to build Sasol II at Secunda; work began in 1976, production in March 1980. In February 1979, as a reaction to the cutoff of Iranian oil as a consequence of the Shah's downfall, the company decided to double the plant's projected capacity to 45,000 barrels per day and to construct a Sasol III facility as well. Construction started that year, and the third conversion plant, also with a capacity of 45,000 barrels per day, began to produce oil in May 1982. Both Sasol II and III received extensive U.S. and West German technological assistance. Sasol had other U.S. connections: It was involved in synthetic fuel projects funded by the U.S. Department of Energy. Connie Mulder, South Africa's minister of information, met with Vice President Gerald Ford and offered to share Sasol knowledge with the United States; Fluor later contracted with Sasol to market its oil-from-coal technology in the United States. Sasol also became a consultant and co-licensor in the development of a coal gasification plant in North Dakota.[9]

The Sasol process—never cost-effective—has been adopted by South Africa out of strategic necessity. Coal is abundant and miners' wages are low, but oil converted from coal has been estimated at $75 per barrel, far more expensive than imported oil.[10] The price differential narrowed during the period 1973-1980 as oil prices soared to over $40 per barrel, but Sasol's cost must now be viewed as exorbitant in light of oil's declining price in recent years. This helps explain why no new Sasol plants have been authorized since 1979, and why no expansion of capacity has been made in existing facilities. In 1981, the chief director of the Department of Mineral and Energy Affairs predicted that Sasol would provide South Africa with 60-70 percent oil dependence in the years 1990-1995,[11] but the upper limit cannot possibly be more than 40 percent. Furthermore, Sasol does not produce much diesel fuel, which necessitates continued high imports of crude for processing in South Africa's refineries. Last, Sasol's role in furthering self-sufficiency has been hampered by strikes at the coal mines supplying the required ore. This internal threat must be factored into any evaluation of the efficacy of external sanctions.

COUNTERMEASURES

South Africa has established a complex state bureaucracy to regulate the oil industry and facilitate the acquisition of fuel by the military and government agencies. The Strategic Fuel Fund (SFF), established in 1964, controlled the strategic stockpile until 1979 and has been the main mechanism for procuring oil. In 1984 the SFF was removed from the control of Sasol and placed under the administration of the Industrial Development Corporation. Also instrumental has been the Strategic Oil Fund, which became the State Oil Fund in 1977 and then the Central Energy Fund in 1985. It financed the construction of Sasol II and III through levies on gasoline sales and now oversees all energy development.[12]

The Equalization Fund, also dependent on a gasoline tax, was initiated in February 1979 to spur oil imports, which increased at an annual rate of only 0.8 percent during the period of 1973-1978 while energy consumption grew by 4.6 percent per year.[13] The fund compensates oil companies for the differential between their cost per barrel and the base price of crude, a differential created by the need to pay a premium to acquire oil by circumvention of the embargo. An $8-per-barrel payment is also given to the companies as a form of subsidization conducive to their maintenance of refineries in South Africa.[14] Premiums were highest during the years 1979-1982, averaging about $8 per barrel, but the oil glut has forced payments down to about one-sixth of that amount.[15] A former minister of economic affairs revealed that the premium was once as high as 70 percent of the base price.[16]

South Africa has been unable to secure oil with its own tankers because they are easily blacklisted. In 1979 the state-run shipping company Safmarine sold its two remaining tankers, the *Kulu* and *Gondwana*, which had sailed under Panamanian flags. In 1984 the government sold its controlling shares of Safmarine.

South Africa has comprehensive laws for securing the compliance of foreign-owned oil companies operating there: These firms must refine oil from any source if they have available capacity. The National Supplies Procurement Act of 1977 requires them to provide a portion of their oil to the government, a stipulation that obviously serves the interests of the military and the police. Refineries must produce those oil products that the government determines are most needed. The National Key Points Act mandates that they provide security for oil installations and permit the military to station personnel there during emergencies. Laws like these force oil companies to cooperate—and give them an excuse to do so.

A lucrative carrot presented to the oil firms is the linkage established between oil imports and coal exports. In 1978 the South African government offered the oil companies coal to market overseas if in return they would assist with the acquisition of crude. The minister of economic affairs, Chris Heunis, declared in regard to coal export quotas that the companies "have been subjected to the condition that they continue to fulfill their obligation in supplying liquid petroleum fuels to the country" and that coal contracts would be "reviewed" if oil deliveries were discontinued.[17] Shell, British Petroleum, and Total were to be given one-third of South Africa's coal exports by 1985; a modification of the arrangement in 1981 raised the quotas for these three companies and provided coal for Agip as well.[18] Though the growing boycott of South African coal in Western Europe could have a deleterious impact on this ingenious crude oil inducement plan, Shell has so far emerged as the world's largest coal trader.[19]

EMERGENCY PREPARATIONS

If oil availability becomes more problematic, South Africa has many means at its disposal to cope with the situation. Advance planning has produced regulations requiring factories to stockpile enough fuel for at least thirteen weeks and sufficient lubricants for one year. Gasoline rationing slips were prepared long ago, in 1973, in case rationing became necessary. Some measures have already been put into practice: Used lubricating oil is recycled, generators providing electrical power have been adapted to save oil, and conversion from oil-fired to coal-fired generators is encouraged through tax incentives. At times of particular shortage, as during the Arab oil embargo of 1973-1974, gasoline stations have been closed on weekends and weeknights; speed limits have been reduced; and bans have been imposed on car racing, waterskiing, and the flying of pleasure aircraft. Extra seats have been added to passenger aircraft.[20] Though gasoline prices have been raised many times, conservation is not the sole motive—surcharges help pay for Sasol plants and compensate for the falling value of the rand.

Maintaining a stockpile of oil as strategic reserves is the key to South Africa's ability to counter the sanctions policy, even though the quantity stored represents unproductive fixed assets that tie up capital. Initial funding for this endeavor was provided in 1964, and oil started to flow into abandoned Transvaal coal mines two years later. The volume stored was increased significantly in late 1977 when the Security Council began invoking an arms embargo and authorities in Pretoria anticipated the imposition of an oil embargo. Oil prices were then rapidly accelerating, which made South Africa's efforts highly expensive.

Based upon the current level of consumption, the strategic reserves can maintain South Africa for approximately two years. But, the reserves can surely last much longer if rationing is implemented or if the supply of bunkers to foreign vessels is terminated. South African sources tend to exaggerate the efficacy of the oil stockpile in order to demonstrate a certain disdain for sanctions. A three-year supply is a frequently cited figure, but there are estimates of up to twelve years based on accompanying conservation measures and increased Sasol productivity. Prime Minister Botha resorted to hyperbole when he stated in 1983 that there were sufficient reserves to withstand a total embargo indefinitely, but there is considerable validity to the argument that the stockpile can buy time for South Africa while a crash effort is made to become self-sufficient in energy.[21]

The reserves were drawn upon extensively in 1973-1974 during the Arab oil embargo, and again from late 1983 to early 1985 to counter rampant inflation and a declining rand. This preferred method for dealing with crisis situations—South Africa has never resorted to rationing or to cutting bunkers—has some disadvantages, however. Early in 1986 the stockpile was replenished by large purchases of crude. Oil prices were falling, but the rand was still weak, and the government in effect opted to use its dollars to buy oil rather than to support the rand.[22]

Another way to deal with oil sanctions is to seek alternative sources of energy. Research is being conducted into ways to substitute sunflower oil for diesel fuel, power vehicles with hydrogen, and extract oil from torbanite (a mineral resembling coal). The South Africans are also striving to increase the amount of oil derived from coal and to develop electric vehicles, especially those used commercially. Either maize or sugar cane may be converted to ethanol and mixed with gasoline or diesel, and gasohol-powered autos running on ten percent ethyl alcohol have been sold since 1980. Nevertheless, methanol is favored over ethanol because it can be derived from coal (a process that is more energy-efficient than Sasol's coal-to-oil conversion).[23] Synthetic fuel usage has been delayed because so much imported oil is available at a low price; in purely economic terms, there is no reason to switch to methanol unless oil costs more than $34 per barrel, and it has been hovering at about half that level in the late 1980s.[24]

THE BOTTOM LINE

Coping with the oil embargo has proven expensive; one South African report in 1986 cited $22 billion as the amount already spent on stockpiling (based on interest loss, storage, and site preparation) and premium payments alone.[25] Once the extra cost of the Sasol process and Soekor's exploration expenditure are factored in, a reliable estimate of the yearly financial impact of the embargo is about $2.3 billion.[26] This figure does not include various hidden costs: loss of export earnings on the coal used by Sasol, synfuel research, losses through fraud, and the payment of commissions and bribes. Although not directly related to oil sanctions, security expenses and damage through sabotage may additionally be viewed as among the costs engendered by economic warfare. Eventually, the expected cost differential of the Mossel Bay gas-to-oil process must also be added to the equation.

Although the cost has been high, South Africa has managed to maintain its energy system at workable levels. Mandated conservation has not been pressed too hard, in part because it could lower the quality of life and, ultimately, cause increased white emigration. Sasol has been at the forefront of Pretoria's counterstrategy, but low oil prices should inhibit new investment in plant capacity for the immediate future. Because nuclear energy provides a viable alternative, it is somewhat surprising that no reactors have been built since the initial one started operations at Koeberg in 1984.

5

STRIKING BACK
Targeting Black Neighbors

South Africa has the capability to cut off oil deliveries to neighboring black states and to organize sabotage raids against their oil facilities because it dominates refinery operations in southern Africa and controls the regional transportation grid, especially the railroads. As early as 1970, Prime Minister John Vorster threatened to use his oil weapon: "Our neighbors are fully aware of the seriousness of the situation and precisely how vulnerable they are should the situation deteriorate."[1] Undeterred, black states continued to press for an oil embargo against South Africa. When the Council of Ministers of the Organization of African Unity met in Addis Ababa in November 1973, oil sanctions against the Pretoria government were strongly endorsed, even though members realized that states like Botswana, Lesotho, and Swaziland could be forced to suffer the consequences. Once Arab exporters had imposed an oil embargo against South Africa, Secretary-General Alfred Nzo of the African National Congress appealed for help to allow these states to withstand possible South African counterpressure.[2]

APPLYING PRESSURE

South Africa exports about 40,000 barrels per day of oil products to southern and south-central African states such as Botswana, Lesotho, Swaziland, Zimbabwe, Zambia, Mozambique, Malawi, Tanzania, Zaire, and the South African–administered territory of Namibia. Botswana, Lesotho, and Swaziland are completely dependent on Pretoria's exports at a combined rate of approximately 6,000 barrels per day, as is Namibia, which consumes at least 10,000 b/d. South Africa controls the Namibian oil supply—it has occupied the former South-West Africa since World War I—and it regulates deliveries to landlocked Lesotho, which is completely surrounded by South African territory. Botswana and Swaziland also lack outlets to the sea, but they border other independent black states and can conceivably switch to alternative oil suppliers: Botswana could import via Zimbabwe or Zambia, and Swaziland via Mozambique. However, these potential providers are themselves partially dependent on South African supplies and would have great difficulty in serving as exporters in their own right, although Mozambique probably has a greater capability than does either Zimbabwe or Zambia.

Botswana, which gets its oil products by rail from South Africa, in 1980 completed construction of storage tanks to provide a buffer in the event of economic warfare directed by Pretoria. In the short term, its effort boomeranged: South Africa refused to permit the transit of oil intended for the tanks, and it was not until 1982 that Botswana was able to fill its strategic reserve, which furnishes a stockpile adequate for about four months.[3]

South Africa has used the oil weapon to interfere in the politics of Lesotho. During a period of poor relations between Pretoria and the government of Leabua Jonathan, Lesotho attempted to secure oil from another source, but South Africa effectively blocked this endeavor. In 1982 Algeria sent oil to be refined in Maputo, Mozambique, and then delivered to Lesotho; South Africa refused to let it be transported through its territory. After eighteen months, Lesotho succumbed and sold the oil.[4] In February 1983 storage tanks in Lesotho's capital, Maseru, were damaged by members of the Pretoria-assisted Lesotho Liberation Army.[5] A December 1985 South African raid on Maseru was followed by a gasoline blockade. The aim—to destabilize Jonathan's government—was achieved on January 20, 1986, when Major-General Justin Lekhanya seized power. That same day, a train with eight gasoline tankers was permitted to cross into Lesotho.[6]

Malawi's South African oil purchases reach the country via tanker to Mozambique and then by rail. At times, the South African-supported Mozambique National Resistance(Renamo) has cut the rail line, even though Malawi lends some assistance to that movement. Also, just before Malawi hosted a November 1981 meeting of the antiapartheid Southern African Development Coordination Conference (SADCC), the Pretoria government registered its disapproval by curtailing oil deliveries.[7]

Zambia has not been the target of pressure, but its economic problems have forced some reliance on South African oil supplies. Zambia generally receives its oil through a pipeline running from Dar es Salaam, Tanzania, and it has a refinery to produce oil products. However, it has suffered from a serious foreign exchange shortage and has had difficulty paying for crude. South Africa has therefore stepped into the breach, providing diesel fuel on credit.[8]

Tanzania tries to avoid any economic interaction with South Africa, but has purchased oil products at times. In October 1983, the tanker *Ardmore* delivered 20,000 tons of diesel and 10,000 tons of jet kerosene from Cape Town. The papers claimed that the fuel had come from Singapore and that the transaction had been worked out through a firm registered in Switzerland. It was unclear if Tanzanian officials realized that the oil products were of South African origin.[9]

Angola is the only regional oil producer. Its need for hard currency leads the MPLA (Movimento Popular de Liberatação de Angola) government to sell its crude to Western capitalist states rather than to its African neighbors, but the South Africans have virtually ensured that Angola cannot easily change course. The country's transportation network and oil facilities have come under frequent attack by South African commandoes and their UNITA (União Nacional para a Independência Total de Angola) allies: Examples include the August 1980 assault on the Lobito oil terminal, the November 1981 damage to storage tanks at Luanda, the July 1984 Cabinda pipeline sabotage, and the unsuccessful May 1985 foray against Chevron installations in Cabinda. The South Africans hope to weaken the MPLA's grip by fostering economic disruption, but a long-range strategic goal could be two gain access to Angola's oil should the UNITA movement ever come to power.

THE ZIMBABWEAN LINKAGE

Zimbabwe, when it was still the white-ruled breakaway colony of Rhodesia, survived Security Council oil sanctions through the supportive efforts of South Africa.[10] Once a black majority government came to power and independence was proclaimed, Zimbabwe hoped to rid itself of this reliance, but most of its aviation fuel and lubricants still arrive by rail from South Africa.

Zimbabwe's predicament has been caused by effective South African pressure on logistics. Because it is technically possible for

Zimbabwe to import almost all of its oil through the pipeline from Beira, Mozambique, the Pretoria government has made a concerted effort to render it inoperative. The Beira pipeline was shut down in 1965 as part of the sanctions directed against the Rhodesian white minority's unilateral declaration of independence. Once Rhodesia became black-ruled Zimbabwe in 1980, efforts were made to reopen the pipeline, but South Africa obstructed these plans by encouraging Renamo sabotage attacks, such as the one in October 1981. The Beira-Mutare (formerly Umtali) pipeline eventually resumed pumping diesel and gasoline in June 1982; crude was not sent through the line because the Feruka refinery in Mutare was still closed. In October 1982 Renamo cut the pipeline and damaged a pumping station. Zimbabwe responded by sending troops into Mozambique to protect its oil lifeline; they were later joined by units from Tanzania, Malawi, and Zambia. Renamo has been striving to obstruct the flow of oil ever since: The most noteworthy efforts have been the blowing up of a fuel storage depot in Beira in December 1982 (South African commandos may have carried out this sabotage) and assaults on the pipeline in January 1983 and August 1985.[11] Constant harassment has had its effect on maintenance also—the pipeline has not been functioning well, and Beira has been plagued by inadequate supportive services such as electricity. In December 1986 diesel generators in Beira suffered from pumping problems, forcing Zimbabwe to purchase additional oil products from South Africa.[12]

The Beira pipeline has become a major focus of the struggle against apartheid as black states attempt to assist Zimbabwe by providing troops; Mozambique has clearly been unable to keep the line open by itself. Mozambique has also been experiencing severe economic problems, but has been aided extensively by Britain in redeveloping the "Beira corridor," the route for the pipeline, a highway, and a railroad connecting Zimbabwe to the Indian Ocean coast. Britain has a historical commercial interest in Zimbabwe, its former colony, and recognizes Mozambique's important role in effecting a transition to black rule there.

Sporadic operation of the Beira pipeline has forced Zimbabwe to import oil by rail from its apartheid antagonist. This gives the Pretoria government some leverage, which it has not failed to use. In mid-1981 diesel deliveries were slowed down; in late 1982 diesel and gasoline cargoes in railroad tankers were temporarily stopped; and from mid-1984 until late 1985 railroad transport was again blocked. Apparent plots to attack oil storage tanks at Beitbridge, just inside Zimbabwe's border with South Africa, were foiled in both January and September 1983.[13]

Zimbabwe has tried to cope with the oil situation by several means. One is increased production of sugar-derived ethanol, which is to be added to gasoline in order to stretch supplies.[14] Another is the plan to build a new oil refinery at Mutare that will be more flexible than the oil Feruka complex in processing different types of crude. Zimbabwe may even seek Angolan crude and offer oil products to Botswana and Zambia.[15] Also, the five major foreign oil firms in South Africa have assured the government of Robert Mugabe that they will provide alternative transport for deliveries should Pretoria impose an embargo against Zimbabwe.[16] Finally, reliance on South Africa for aviation fuel has been reduced since the Beira pipeline began pumping it to Mutare in 1985.[17] Despite all of these programs, Zimbabwe's geographical location tends to foster dependence on South Africa, and crucially needed oil products still travel northward up the rail lines irrespective of Zimbabwe's strong public denunciation of apartheid and its support for sanctions against South Africa.

South Africa's warnings and pressure have only been used for tactical, regional reasons, not to counteract the embargo. South Africa undoubtedly could make good on its threats if faced with an oil crisis, but would be undercutting its own position—oil exports to other African states are profitable and also keep South Africa's refineries operating closer to capacity. In addition, use of the oil weapon would prompt the black states to arrange alternative sources of supply, and assistance could probably be expected from Nigeria and some non-African producers such as Iran. Logistically, the main problem would be transporting oil to landlocked Lesotho—an airlift over South African territory would be required.

6

THE LEAKY SIEVE
Circumventing
Oil Sanctions

South Africa was able to circumvent attempts to apply an oil embargo thanks to ample quantities of crude from Iran. But growing instability there as Islamic fundamentalists came to power presented a major challenge to Pretoria, forcing reorganization of its oil import system. Overcoming the oil crisis of 1979 was a multifaceted undertaking carried out with considerable ingenuity and created the framework by which South Africa has coped with oil sanctions ever since.

MINOR INCONVENIENCE

Prior to the 1973-1974 Arab oil embargo, the largest suppliers to South Africa were Iran, Iraq, Saudi Arabia, and Qatar.[1] Cutoff of Arab deliveries in November 1973 caused crude imports to drop from 12.4 million tons in 1972 to 11.7 million tons in 1973.[2] Pretoria's reactions was to ban gasoline sales on Sundays, reduce speed limits, outlaw the use of fuel for some sporting events and to terminate the publication of oil statistics.[3] The Arab embargo was imposed on South Africa just when summertime gasoline use was about to peak. Also problematic was the shortage of light-aircraft fuel, a product South African refineries were unable to produce; a tanker delivering needed supplies broke down in November[4].

The embargo proved only a minor irritation because counteraction quickly followed. Oil imports jumped to 13.6 million tons in 1974, although rapidly rising prices increased the cost more than fourfold over 1973 deliveries.[5] Iran stepped into the breach by furnishing 87 percent of South Africa's crude imports for 1974-1975, and the end of the Arab embargo in March 1974 led to renewed shipments from Iraq, Qatar, and the United Arab Emirates. Brunei also started to become an important supplier. When the Arab embargo was imposed, Iran's minister of finance, Jamshid Amouzegar, had assured South Africa that his country would neither halt nor reduce supplies—Teheran more than fulfilled this commitment.[6]

Iran had a special relationship with South Africa. Shah Mohammed Reza Pahlevi's father, Reza Shah, sought refuge there after his 1941 abdication and was buried in Johannesburg following his death in 1944. In the oil sector, the National Iranian Oil Company owned 17.5 percent of the Natref refinery, which Iran helped construct, and had a contract to supply it for twenty years once operations began in 1971; and Iranians were granted the status of honorary whites according to South Africa's convoluted racial designations.[7] By the time of the Arab embargo, Iran was already furnishing almost 60 percent of the oil processed by the Natref refinery.

Iran, which had no interest in curtailing deliveries to South Africa, had tried to deflect blame by claiming that foreign oil companies, not the government, were primarily responsible for supplying South Africa. Iran indicated that it would participate in an embargo only if all other countries did likewise and the Security Council mandated sanctions, an unlikely possibility at the time. When the General Assembly first voted to implement oil sanctions in 1963 in conjunction with steps against South Africa's occupation of Namibia, Iran joined with the majority, at the same time announcing that its vote did not mean any new sanctions would be applied. As none were then operative, Iran's vote rang hollow, and the Shah's regime continued to be the main supplier of crude.[8] Iran then increased deliveries to compensate for the Arab embargo of 1973-1974, remembering well that Arab states had moved into its markets while Iran's oil was being blacklisted during the 1951-1953 nationalization dispute with the British. During the period 1973-1978, Iran provided 90 percent of South Africa's crude imports; the figure rose to 96 percent by late 1978.[9] Most of this oil was supplied by companies belonging to the Consortium, a group of foreign companies selling Iranian oil, rather than by the National Iranian Oil Company directly.

FALL OF THE SHAH

Prior to Iran's revolutionary upsurge of 1978, there was some indication that its oil ties to South Africa were producing strains in both countries. Some South Africans, concerned about overreliance on Iran because of its advocacy of higher oil prices, expressed fear that the Shah's successor would not be as favorable to Pretoria's interests.[10] Iran, to reinforce its image as a Third World state, informed the South Africans that its oil could not be reexported to Rhodesia. When Foreign Minister Roelof (Pik) Botha visited Iran in October 1978, where turmoil had already erupted, he was told that Iran would embargo deliveries if South Africa imposed a unilateral settlement in Namibia.[11]

It was the Iranian revolution—not sanctions—that had the major impact on the South Africa oil situation. Strikes by Iranian workers began in September 1978, with a general strike of oil workers initiated on December 4; production fell to less than a fourth. Ayatollah Khomeini, from French exile, first called for a ban on all oil exports on November 23 in an effort to cripple the Shah's economy, and on December 13 he warned that states supporting the Shah would get no oil once his forces were in power.[12] It became difficult for oil companies to juggle their supplies in an effort to fuel South Africa because the loss of almost 5 million barrels per day of Iran's production had severely tightened the world oil market. British Petroleum, a major provider of oil for the Pretoria government and the company most reliant on Iranian production (it drew 40 percent of the crude made available to the Consortium), was hit especially hard and was unable to fulfill its South African contracts.

In the midst of a revolutionary crisis, the Shah on January 4, 1979, appointed Shapour Bakhtiar as prime minister. On January 11, five days before the Shah left for exile, Bakhtiar announced that oil would not be furnished to South Africa, although it appears that strikes had prevented any such shipment since the last week of December.[13] On January 31 Ayatollah Khomeini returned to Teheran, Bakhtiar resigned on February 11, and the new Islamic administration vowed to embargo South Africa. As we shall see, the words were not quite matched by concrete action.

RISING TO THE OCCASION

The Iranian revolution precipitated a 40 percent drop in South

Africa's crude imports for the first quarter of 1979 as compared with the corresponding period a year earlier. By June the supply was still 30 percent below the December 1978 level, and imports for the year fell by 25 percent.[14] Nevertheless, South Africa survived the crisis. Aided by the newly created Equalization Fund, it turned to the spot market, where it offered lucrative premiums for cargoes of crude. Booming prices on gold, platinum, and diamond exports helped pay the bill, but there is no concrete evidence that gold was actually bartered for Saudi or any other oil.[15] At first, spot market loads provided a stopgap, with crude purchased through the Rotterdam market and elsewhere. According to a South African source, two tankers laden with oil from South America were already on their way in January 1979 to replace the Iranian supply.[16] Gradually, longterm contracts were worked out to assure Pretoria of dependable deliveries of crude.

The limited quantity of oil available necessitated conservation measures, especially in the transport sector. Fuel price increases were applied in February 1979 and again in June, while speed limits were reduced and gasoline stations closed on Saturdays.[17] Because of the tight supply, no oil was added to the strategic reserve. On the other side of the ledger, efforts were accelerated to provide alternative sources of energy. A decision was made to expand Sasol II, which was not yet operating, and to build a third Sasol plant. For the short term, however, Sasol was not an important factor; it produced only 7 percent of South Africa's oil. Some attention was given to the idea of using sunflower oil as a substitute for diesel, with the minister of agriculture calling upon farmers to devote 10 percent of their land to growing sunflowers.[18] In addition, the search for offshore oil was stepped up, though without success.

By September South Africa had reorganized its oil sector, and the Iranian revolution no longer had a significant impact. Indicative of Pretoria's newfound confidence was the lifting of the Saturday gasoline ban and the raising of speed limits. Even more reason for pride was that the crisis had been overcome without having to resort to the strategic reserve; the chairman of Sasol declared that not "a single drop" had been utilized.[19] Furthermore, the rationing coupons for gasoline had gone unused, and the flow of oil to Rhodesia had continued undiminished. South Africa had clearly withstood its ordeal and could now bask in the glow of soaring gold prices, which peaked at $875 per ounce in January 1980. But a new problem developed: The rand began to decline in value. Roughly equal to the dollar in the early 1980s, it slid to about 57 cents in 1985 and then to less than 45 cents in 1986. Gasoline prices had to be increased, with raises of 40 percent and 10 percent taking place in February and November 1985 respectively. Fortunately for the South African economy, the price of oil was dropping along with the value of the rand, so the strain engendered by weak purchasing power never became too severe.

UNEASY PARTNERS

Some diplomatic and economic interaction continued between Iran and South Africa despite public denial. Iran officially broke relations on March 4, 1979, and asked the South African consul-general to leave, and the Islamic government hosted a delegation from the African National Congress in November of that year. In practice, relations were continued at the consular-general level, although the diplomatic representatives' rank was reduced to that of acting consul-general,[20] and a South African consulate therefore operated unofficially in Teheran. Trade was not actually curtailed as most observers were led to believe: South African steel tubing and timber were still shipped to Iran, as were industrial plastics falsely listed as having originated in Swaziland or Mozambique.[21]

Iran continued to furnish South Africa with crude despite its announced embargo, although deliveries were infrequent and Teheran was no longer the primary source of supply. Even at the height of the 1979 crisis, two tankers brought Iranian oil to Durban. A third, the Shell tanker *Latona*, had started to unload in the United States, but was diverted to Cape Town with half its cargo remaining on board.[22] If Iranian authorities were unaware that these crude oil shipments were being routed to South Africa, they must have known of the extensive deliveries that followed. During the period 1979 through August 1987, at least 36 Iranian cargoes totaling more than 9 million tons arrived in South Africa.[23] Teheran also maintains its 17.5 percent share in the Natref refinery, but according to the Iranians, they have been excluded from Natref board meetings and efforts to sell their percentage of the refinery have been blocked by Sasol.[24]

It also appears that Iran has bartered oil for arms needed to pursue the war with Iraq: Shirebu has identified three ships transporting military equipment furnished by the South Africans during the period 1984-1985.[25] Allegations have been made concerning military shipments via the Comoros and a $30 million arms-for-oil agreement in which a British dealer was to provide antitank missiles.[26] There have also been reports that the South Africans have worked through a Greek firm to trade arms for oil. The Hellenic Explosives and Ammunition Industry (Elviemek) is reported to have sent Western arms to Iran in return for oil deliveries to South Africa, and the former president of the Greece–South Africa Association is believed to have bought into the ownership of Elviemek in 1984. The following year, a South African resident originally from Greece acquired 30 percent of the company and gained leverage over the remaining 70 percent through a firm under his control.[27]

Overall, the Iranian role in the South African oil sector is smaller than it was prior to the revolution. But that Iran is involved at all belies its public claims about enforcing an embargo and its official pronouncements about the odious nature of apartheid. Though Iran has promised to blacklist companies found to be delivering crude to South Africa, there is no evidence to corroborate that this policy is being carried out.[28]

7

MAINTAINING THE FLOW
Securing Oil Supplies

South Africa's reliance on Arab oil resumed after the Iranian revolution forced Pretoria to turn to the spot market in 1979. Arab oil accounted for perhaps one-third of South Africa's imports that year and in 1980, as Pretoria worked out long-term contracts with Saudi Arabia and Oman. Throughout the 1980s Arab crude has found its way to South Africa despite official embargoes verbally supported by all Arab producers. The oil glut encourages sales: Iran, Brunei, and even some black African states have participated behind the scenes.

CLANDESTINE ARAB CRUDE

Saudi Arabia, Oman, and the United Arab Emirates are among the four major crude suppliers of South Africa (the other is non-Arab Iran); Qatar and Kuwait provide oil as well. Occasional shipments come from Egypt and Libya, but deliveries of Iraqi crude were largely curtailed by the Gulf war.[1] According to Shirebu's detailed reports, most of South Africa's crude comes from Arab states in the Gulf: Twenty-eight of the 52 ships identified as likely to have delivered oil to South Africa during the period January 1980–June 1981, sailed from the Gulf. From July 1981 through December 1982, the figure was 40 of the 57 apparent deliveries, and for the years 1983–1984, 64 out of 83. During this last period, an estimated 15.5 million tons of crude reached South Africa on voyages that could be identified, and the Gulf states furnished 13.6 million tons of this total. Iran is also a Gulf exporter, but only 7 of the cargoes were believed to have originated there.[2]

Saudi Arabia claims to adhere to an embargo, but has officially stated: "The Government of the kingdom of Saudi Arabia does not maintain a list of oil companies or tanker companies that have violated the contracts of sale or shipping by supplying or shipping oil and petroleum products to South Africa."[3] The absence of a blacklist of offending companies makes it easier for Saudi Arabia to look the other way if its crude is discharged in South Africa, and it is therefore not surprising that Saudi Arabia is South Africa's biggest supplier: Shirebu has reported at least 112 Saudi deliveries from January 1979 through March 1985, the 112 cargoes totaling 22 million tons; this represented more than a quarter of Pretoria's crude imports.[4] Deliveries have tapered off somewhat in recent years. The groundwork for such extensive trade was developed after the Iranian revolution, and in December 1979 Sasol's London manager reportedly met in Saudi Arabia with the director of Petromin, the state oil company.[5] In June 1984 South Africa's advocate-general issued a report on his investigation into aspects of the oil trade—the "Z Country" cited as a main supplier has been identified by an analyst of the text as Saudi Arabia.[6] Saudi-South African oil relations were so cooperative that Pretoria's minister of justice, Jimmy Kruger, became a broker of Saudi oil in the U.S. market after he left office, earning a quarter of a million rands for a deal completed in 1982.[7]

An interesting angle to Saudi-South African oil ties has been provided by California businessman Sam Bamieh. A naturalized U.S. citizen of Palestinian origin, Bamieh testified before a congressional committee that Prince Bandar bin Sultan, the Saudi ambassador to the United States, had made certain proposals to him at a meeting in Cannes, France, in February 1984. According to Bamieh, Prince Bandar asked him to set up a company that would channel material support to anticommunist forces in Angola, Afghanistan, and Central America. He also wanted Bamieh to act as a middleman for the sale of Saudi crude to South Africa, offering him a profit of 75 cents to one dollar per barrel to do so. When Bamieh displayed reluctance, Prince Bandar allegedly tried to reassure him that he would be serving U.S. interests because CIA director William Casey and King Fahd were at that very time aboard a yacht in the Mediterranean discussing the same topic. Bamieh testified he had no personal knowledge of this Casey–Fahd conclave and was relying on remarks made to him by Prince Bandar. Fearing he would be drawn into illegal activities, Bamieh turned down the Saudis, but he believed that others stepped in to contract for the oil during the period December 1985 to February 1986.[8]

Bamieh's testimony touched upon Saudi aid to UNITA in Angola and U.S. provision of AWAC aircraft to Saudi Arabia. Though he did not directly link these issues, inferences have been drawn—and perhaps exaggerated—by some observers.[9] First, it is important to realize that the existence of the Casey-Fahd meeting has not been adequately verified, so any deal worked out there is highly conjectural. Secondly, Bamieh was engaged in legal and financial disputes with the Saudi royal family prior to giving his congressional testimony, so his implication of the Saudis in various activities could possibly be construed as prejudiced. Nevertheless, it does appear that an arrangement was worked out whereby the Saudis, in return for U.S. military deliveries (especially AWAC), sent weapons to UNITA, the Nicaraguan contras, the Afghan mujahedin, and perhaps Renamo in Mozambique. In addition, although the evidence is inconclusive, the Saudis may have increased oil shipments to South Africa to please Casey and may have accepted South Africa's transfer of military equipment to UNITA in lieu of payment for some of the crude.

Israel, which has frequently been condemned for its trade and military contacts with South Africa, feels that Arab relations with the Pretoria government have not been sufficiently exposed. Even Shirebu in its early delivery reports was reluctant to identify specific Arab states, instead tending to pressure Western oil and tanker companies and governments. (Shirebu also favored the term "Arabian Gulf" to "Persian Gulf.") Shirebu has become more diligent in citing Arab circumvention of the oil embargo in its reports, which the Israelis have publicized in order to demonstrate that Arab economic ties to South Africa far exceed Israel's. As the Israeli representative at a 1986 trade union conference indicated, Arab states were furnishing three-quarters of Pretoria's oil imports, yet Israel was purchasing less than one percent of South Africa's exports and providing only 0.75 percent of its imports. Israeli investment represented only 0.1 percent of the external total.[10]

In 1984 an Israeli lobbying group in the United States, AIPAC (American Israel Public Affairs Committee), released a report on Arab oil deliveries to South Africa. Using Shirebu figures as a basis for further investigation through the Lloyd's Voyage Records, AIPAC concluded that South Africa received 76 percent of its oil shipments from Arab states (worth about $2.3 billion per year) and 6 percent from Iran from mid-1981 through 1982. The leading Arab suppliers were Saudi Arabia, United Arab Emirates, and Oman, and the import figure had risen from only 38 percent of shipments at the beginning of the 1980s.[11] In 1986 Israel issued a study covering the period 1980–1984. The methodology was similar to that used in the AIPAC report, but there were two serious flaws: Non-Arab Iran was not isolated from statistics referring to the Gulf, and tankers unloading in South Africa and then proceeding to the Gulf were inexplicably included. Therefore, listed were voyages to South Africa from South America, Western Europe, West Africa, Brunei, and the United States. But many of these vessels certainly did not carry Arab crude, and thus the interpretation as to the Arab role is misleading—though Arab states did permit these tankers to dock despite their having come directly from South Africa. Of the 167 tanker total in the Israeli list of voyages, 41 apparently do not represent voyages from Arab states to South Africa, and 31 are questionable because of multiporting or the designation of the Gulf, which could in some cases be Iran.[12]

ASIAN ASSET

Brunei in Southeast Asia has been an important source of crude for South Africa since the Iranian revolution. Brunei Shell Petroleum, jointly owned by Royal Dutch Shell and the government of Brunei, has been the major supplier of crude, and British Petroleum has played a role as well. As early as 1979, Sasol had a contract with Shell to supply the Natref refinery with oil from Brunei.[13] According to a Shirebu report: "It is unlikely that Shell was not conniving at the continuing crude oil deliveries to South Africa from Brunei. Especially so, because the company operates on both ends of the contract: the sale of Brunei crude oil, and the import of this embargoed crude oil into South Africa."[14] Shell maintained that the embargo was not properly instituted because states favoring its application lacked enabling legislation and oil companies such as Shell had not been adequately informed as to the embargo guidelines. Consequently, Shell believed itself entitled to deliver oil to South Africa.[15] By 1981 Shell stopped furnishing the oil directly, but Brunei Shell Petroleum began to sell it to a Japanese company that in turn arranged for deliveries to South Africa through a U.S. broker, Marc Rich.[16]

Shirebu has identified 56 deliveries from Brunei to South Africa during the period January 1979–October 1986 totaling about 6.7 million tons of crude, roughly 6 percent of Pretoria's imports over that span.[17] Brunei was unusual in openly selling oil to South

Africa, and its endorsement of the embargo in 1982 did not produce any discernible change of policy. But Brunei was still a British protectorate at the time (until January 1, 1984), so London must bear some of the responsibility for fueling South Africa. (The former British colony of Singapore is another example; Neptune Orient Lines Ltd., 70 percent government-owned, is alleged to have transported six cargoes during the years 1983-1984.)[18] Brunei, sensitive to criticism of its role as an oil supplier, agreed to investigate whether thirty cargoes of its crude sold to a Japanese company were possibly forwarded to South Africa. Its trade statistics could also bear scrutiny: Shirebu has reported a large discrepancy between the amount listed for exports to the United States and the quantity the United States claims it imports from Brunei.[19]

A CONTINENTAL AFFAIR

Several black African states that strongly condemn apartheid have nonetheless been active in supplying South Africa with oil. Mozambique sold heavy fuel oil, as well as partially refined oil that it could not process. As of 1981, it may have been furnishing more than 40 percent of the production of the Petromoc refinery in Maputo, about 3,600 barrels per day. Apparently, a Liberian flag tanker under charter to Socal began carrying the oil to the Caltex refinery in Cape Town in January 1980, with deliveries made about every three months thereafter. Some of the fuel oil produced by Caltex was then sold back to Mozambique because the Petromoc refinery was incapable of turning out certain products.[20]

A prominent Pakistani commercial family is alleged to be operating several companies engaged in providing oil from black African states, and a firm linked to South Africa's Anglo American Corporation is believed to be responsible for a cargo of partially refined oil delivered from Madagascar in July 1982.[21] There have also been rumors that Pretoria has been acquiring oil with the assistance of the Comoros, Equatorial Guinea, the Seychelles, and Cameroun.

Nigeria, the most prolific oil producer in black Africa, may also be a source. Officially, it has played a leading role in imposing sanctions: In 1978 all foreign companies operating there were required to list their connections to South Africa; public money was also withdrawn from Barclays Bank of Nigeria to protest its parent company's links to the apartheid government. In 1979 ships with

South African ownership, passengers, or crews were banned from Nigerian ports, as were vessels that had called at South African ports during the previous six weeks. In 1982 the Nigerian government directed its federal and state departments not to deal with companies tied to South Africa and required these companies to declare any business dealings with Pretoria.[22]

Unofficially, oil has slipped through the antiapartheid net and Nigeria has acknowledged that it is faced with a situation difficult to control. Fraudulent shipping documents, including bills of lading, are readily available in Nigerian ports—a court in 1984 sentenced to death the Spanish master of a Panamanian tanker and two Nigerians for illegally exporting oil.[23] During the early 1970s the government already suspected that Nigerian businessmen were supplying oil to South Africa, and a task force was appointed in 1977 to investigate the smuggling of oil to neighboring states (South Africa was not specifically mentioned). Little headway was made, and the practice of illegal exports continued. A study of voyage records by the International Oil Working Group gave reason to believe that many cargoes found their way to South Africa during the period 1976–1980, and Nigeria's oil minister later revealed that over the years 1979–1983 crude was transported from Nigeria on 189 tankers "to destinations which they were not supposed to reach."[24]

Nigeria is still looking into this vexing problem. Its government does not appear to be involved in transactions with South Africa, but it has been unable to put its economic house in order. Corruption and clandestine business arrangements underlie the South African oil trade, and the use of foreign middlemen makes it even more difficult for the Nigerian government to assert control.[25] One case in which Nigeria was able to act decisively involved the 220,300-ton Liberian-flag tanker *Jumbo Pioneer*. Authorities prevented it from loading oil at Brass in March 1979 because they feared the cargo was intended for Cape Town. The actual ownership of the tanker was unclear, although it was listed as the property of the Cyrus Tanker Corporation, but armed Israelis were aboard when it docked. The vessel was seized, though soon released.[26]

Another interesting episode shortly thereafter featured the Panamanian-flag tanker *Kulu*, which arrived off Bonny on April 30 and soon was loaded with 213,000 tons of crude. Officially owned by a company in Bermuda, the tanker actually belonged to the South African merchant carrier Safmarine. At the time, it was chartered to British Petroleum. The *Kulu* was seized and taken to the Nigerian capital, Lagos, when 27 South African officers and crew were found aboard, along with 12 British and 2 Dutch subjects. Nigeria, infuriated that South Africans were attempting to secure oil there, sold the *Kulu*'s cargo for approximately $30 million without compensating BP. Though the ship was then released to BP, extensive retaliation measures against the British oil company soon followed. The *Kulu* affair demonstrated Nigeria's sensitivity to its unintended underground oil relationship with South Africa, but evidence indicates that the cargo was bound for Rotterdam and Wilhelmshaven in Western Europe rather than for South Africa. In any case, the tanker was indeed South African.[27]

8

LOOKING WESTWARD
Shippers and Middlemen
Assisting South Africa

Most of South Africa's imported crude comes from the Gulf, but the oil and shipping companies participating in this trade are mainly Western. Particularly noteworthy among the circumventors of the embargo have been firms from the Netherlands, Norway, and Britain; those from West Germany and Denmark have also played significant roles. U.S. companies, although prominent in the South African oil sector, have been of only marginal importance in providing Pretoria with crude.

Western states often condemn apartheid, but their economic and strategic interests lead them to apply sanctions loosely. Their oil and shipping interests are taken into account, as is South Africa's ability to provide not only strategic minerals but bunkers, which are of utmost importance along the Cape sea route. South Africa's crucial location and well-developed harbor facilities received a major boost when the Suez Canal closure from 1967 to 1975 led to increased tanker traffic from the Gulf to Western Europe and the United States via the circum-Africa route. In addition, modern supertankers are unable to use the Suez Canal because of the waterway's limited draft. Thus practical considerations have come to supersede an apparent moral imperative.

THE DUTCH CONNECTION

Sanctions instituted by Western states are usually weaker than their official pronouncements may indicate. In the Netherlands, the government calls for a mandatory Security Council oil embargo, and there is strong public activism against apartheid. Nevertheless, deliveries to South Africa are not banned, Rotterdam serves as an important transshipment point, and Dutch oil companies have been active in supplying Pretoria with crude.

Parliament pushed for effective sanctions in November 1979 by voting to impose a unilateral embargo if agreement to coordinate with the European Economic Community could not be reached. When the EEC negotiations stalled, Parliament in June 1980 again endorsed an embargo; the government refused to comply with its decision, and a vote of no confidence was taken. The government survived by a two-vote margin, but its position since has been that sanctions must be coordinated with at least Belgium and Luxembourg because of obligations under the Benelux Treaty. Though the Netherlands has avoided direct responsibility, the government's support for voluntary sanctions by Dutch companies has produced a somewhat chilling effect.[1]

During the period January 1980–June 1981, Shirebu identified 52 cargoes as likely to have been delivered to South Africa. Of the 30 cargoes for which ownership was known, all but four were owned by Dutch companies. Prominent Dutch oil broker John Deuss, operating through several firms, was Pretoria's major supplier at the time, and Royal Dutch Shell was active until 1983, delivering about 5.2 million tons over the years 1979–1982. Shell also owned or chartered many of the tankers plying the South African oil trade.[2] Rotterdam served as a crucial port of origin for the oil transported to South Africa, accounting for almost 10 percent of the crude delivered in 1979–1980. Since February 1981, no tanker leaving Rotterdam has listed South Africa as its destination, but Shirebu investigations have shown that cargoes have arrived there. However, the use of Rotterdam as a transshipment center has declined greatly since 1982.[3]

The Dutch play another role in facilitating the flow of oil to South Africa, this one through the actions of the Netherlands Antilles, a Dutch possession. The situation is analogous to the Britain-Brunei relationship prior to January 1984, in which Shell was also involved. However, unlike Brunei, the Netherlands Antilles does not produce crude—it only transships oil received from Venezuela, Mexico, Nigeria, and other countries. From January 1979 through March 1980, at least 19 tankers carried cargoes from the Netherlands Antilles ports of Curaçao, Aruba, and Bonaire to South Africa. Eight of these vessels were owned by Shell. Furthermore, 7 tankers discharging oil in South Africa during 1980 and the first half of 1981 flew the flag of the Netherlands

Antilles.[4] As in the case of the Netherlands itself, the importance of the Netherlands Antilles in breaking oil sanctions then declined.

THE NORSE SAGA

Norwegian tankers have been instrumental in delivering oil to South Africa. Norway has only 7 percent of the world's tanker tonnage, but its vessels accounted for 35 percent of Pretoria's imports in 1981–1982 and 25 percent over the period 1979–1982. Of the 103 tankers identified by Shirebu from January 1980 through December 1982, 65 were Norwegian. This prominent transport role continued in 1983–1984; 51 of the 83 identified tankers delivering to South Africa were owned or managed by Norwegian companies, and these vessels supplied approximately 40 percent of Pretoria's needs.[5] One company alone may have furnished 7 percent of South Africa's supply over the six years beginning January 1979, and a prominent Norwegian shipper active in deliveries to South Africa was arrested on fraud charges in June 1986 for allegedly using part of an oil cargo for the ship's bunkers.[6] Some Norwegian tankers operated as veritable shuttles: The *Staland* made 8 deliveries over 1981–1982, the *Thorsholm* at least 7 from April 1982 through September 1984, the *Norse King* at least 6 during the months May 1980–September 1981, and the *Havdrott* at least 13 over a three-year period starting in mid-1979.[7]

Gradually, the Norwegian government began to take steps to apply shipping sanctions. In 1985 the minister of commerce and shipping traveled to Greece, Liberia, and Panama in an effort to coordinate policies, but he met with little success.[8] His government, over the objections of the powerful Norwegian shippers' lobby, then developed a system based on a voluntary ban on tankers going to South Africa and the registration of noncomplying vessels. The Storting (parliament) wanted compulsory registration backed by penalties, but a looser form of sanctions was applied by the government in cooperation with the Norwegian Shipowners' Association. Effective April 1, 1986, tankers supplying South Africa were required to register—but the shippers were to step forth voluntarily, and the procedure was to be supervised by the shipowners themselves. Visits to South Africa were to be listed in terms of frequency and cargo load, but the names of tankers and their companies were not to be made public. Though the policy did not terminate the activities of Norwegian shippers supplying South Africa with crude, it represented a small first step against apartheid.

During the first three months of the registration system, four Norwegian-owned ships (only two of which flew the Norwegian flag) were recorded as having delivered 925,000 tons of crude.[9] Obviously, the Norwegian role still loomed large—although it appears the registry failed to include all voyages by Norwegian ships: For example, the March 1987 delivery from Iran by the *Berge Princess* technically did not qualify as a Norwegian tanker; flying the Liberian flag, it was sold to a company in Lichtenstein and then chartered back to Norwegian interests. It therefore was not under Norwegian ownership or sailing under the Norwegian flag.[10]

In March 1987 Norway strengthened sanctions (effective in July) by banning oil deliveries on Norwegian-flag vessels or on ships controlled by Norwegian companies. However, many loopholes remained: The basic governing principle (originally formulated in 1984) was the shipper's intent at the time of contract; if the plan was to supply crude to South Africa, the voyage would be illegal. But the policy could not effectively be applied to time charters. If a Norwegian tanker was chartered out to a company for two years, the intent of each voyage could not possibly be known at the time of the original contract. The same holds true for cargoes rerouted at sea, a common practice in the oil trade, or for cargoes resold at sea and then sent to a different destination.[11] Norway's tanker sanctions thus lack teeth, but it must be realized that no controlling legislation at all existed until 1986.

Norway initiated an unofficial ban on the delivery of its North Sea crude in 1979, but some still managed to reach South Africa. There were at least five cargoes during the period 1979–1982, most notable being those aboard the tankers *Spey Bridge* and *Cast Fulmar*.[12] The *Spey Bridge* left Teesside, England, in December 1980 with over 100,000 tons of Norwegian crude and then unloaded in Durban despite its French discharge certificate, apparently fraudulent. In April 1981, 30,000 tons of Norwegian oil was sent to Rotterdam for storage, then mixed with other crude aboard the *Cast Fulmar* and delivered to Durban in May.[13]

The best-known Norwegian cargo not reaching South Africa was aboard the *Jane Stove*, a Norwegian-owned tanker that departed Teesside just five days after the *Spey Bridge*. Loaded with 125,000 tons of crude, it supposedly was sailing to Curaçao, but was actually headed for South Africa. When a Norwegian newspaper reported the voyage, the government ordered the tanker not to discharge in Durban, and the cargo was eventually delivered to Japan. Ironically, the owner of the vessel was then serving as the chairman of the Norwegian Shipowners' Association.[14] The *Jane Stove* affair precipitated a meeting called by the minister of oil and energy, who stressed to the heads of the sixteen oil companies operating in the Norwegian sector of the North Sea that selling crude to South Africa was contrary to government policy.[15] In June 1986 this policy was formalized, and now expatriate companies purchasing Norwegian crude must certify in writing that it will not be supplied to South Africa. With this policy and the 1985 ban on exports of oil products, Norway has greatly tightened sanctions in regard to both tankers and crude.

THE BRITISH BROUHAHA

Britain has served South Africa's oil interests in many ways. Shell and British Petroleum operate in South Africa, both previously supplied oil directly, and Brunei was permitted to circumvent sanctions when still a British protectorate. In addition, Britain has not imposed a ban on the export of oil products or on the use of its tankers, and crude oil has been delivered even as political controversy over the subject continues.

Shell tankers were active in supplying South Africa with crude, much of it from Oman, until mid-1981; estimates are that at least 23 deliveries were made from February 1979 through January 1981 totaling approximately 4.5 million tons,[16] after which Shell shifted to intermediaries. For its part, British Petroleum played only a minor role, with only 4 reported cases of deliveries over the period 1979–1982.[17]

As has been mentioned, Norwegian oil from the North Sea has occasionally been pumped into tankers at Teesside and sent to South Africa, and 6 cargoes of British North Sea crude as well appear to have been transported from January 1979 through March 1980.[18] Officially, there was no British ban, and this issue became the matter of considerable debate once the Conservatives rose to power in May 1979. Under the previous Labour government, it had been unannounced policy to prohibit the export to South Africa of British crude by restricting sales to members of the European Economic Community and "traditional customers." Permission had also been denied for an indirect arrangement whereby BP would supply North Sea oil to the Continental Oil Company (Conoco) of the United States, and Conoco would then provide South Africa with an equal quantity available in its stocks elsewhere.[19] The new Conservative foreign secretary, Lord Peter

Carrington, chose to reverse this policy and told BP that swap deals with Conoco would be allowed. Consequently, BP furnished crude for Conoco's Humberside refinery, and Conoco supplied South Africa with oil from Indonesia.[20] Lord Carrington explained that Britain did not favor imposing an embargo against South Africa and that "the guidelines on exports of North Sea oil do not have legal force and that there can be no absolute certainty about the final destination of any oil sold to third parties.[21]

The Labourites had tried to head off this move. Outgoing Foreign Secretary David Owen wrote to Lord Carrington indicating that the previous government had rejected the same swap arrangements: "In view of the present acute shortages for British consumers, I would be grateful for your assurance that no North Sea oil has passed to South Africa in any way, either directly or indirectly, and that you have no intention of allowing either that specific application from BP or similar applications in the future."[22] Though Lord Carrington approved BP's request, his reply letter agreed with Owen's position that British crude should only be furnished to members of the EEC and the International Energy Agency (plus Finland). He stated, however, that BP could exchange North Sea crude "for non-embargoed third country crude which can be supplied to their South African subsidiary." He added: "To Prevent the companies from making such arrangements to meet their contractual obligations would be to seek to impose an oil embargo on South Africa, which is not our policy nor, as I understand it, that of the previous Government."[23]

The political tumult in Britain was quickly overshadowed by the reaction in Nigeria. Already at odds with BP over the *Kulu* affair that May, Nigeria had declared a 25 percent reduction in the crude to be supplied to BP effective August 1. Nigeria was also concerned about BP's role in violating the embargo against Rhodesia and about Britain's expected lifting of sanctions against the breakaway colony. News of the Conoco swap arrangement with BP (owned 51 percent by the British government) fueled the controversy—effective midnight July 31, 1979, Nigeria nationalized BP. It was no longer permitted to lift Nigerian crude or to operate its shares of exploration and marketing companies there.[24]

Nigeria did not claim that BP was providing its oil to South Africa—only that its own sales to BP permitted the British company to free other oil for the South African market. Nigeria therefore saw its action against BP as a function of applying sanctions to South Africa, and it was no coincidence that the announcement regarding nationalization came on the eve of a Commonwealth conference in

Lusaka, Zambia.[25] The nationalization cut off 10 percent of BP's crude supply and brought an end to the brief arrangement with Conoco. Economically, Nigeria fared well because the spot market price of crude was then higher than the long-term contractual amount paid by BP. In 1981 Nigeria and BP resolved the nationalization issue, with BP receiving 60 million pounds in crude and the right to resume operations in Nigeria.[26]

Britain, still under Conservative rule, generally prevents its crude from being supplied to South Africa, but its position on sanctions is weak. Oil products are sold, and no controls are placed on the actions of British oil companies or tankers. This posture in the oil sphere stems from the Thatcher government's belief that economic pressure is not the appropriate means by which to spur significant changes in the apartheid system.

THE SUPPORTING CAST

West Germany, Denmark, and the United States have contributed significantly to the flow of oil to South Africa. The West German firm Marimpex has supplied numerous cargoes of crude, many of them leaving from Rotterdam. The tankers involved never fly the West German flag, and they list misleading destinations such as Bonaire, Bombay, Gibraltar, or Bilbao. Shirebu has estimated that at least 10 Marimpex cargoes were discharged in South Africa from December 1980 through May 1983 and that Marimpex supplied about 2.4 million tons of crude in 1983–1984 alone.[27]

Denmark's main contribution to South Africa's oil sector has been through tanker transport. One company, A. P. Moller, accounted for more shipments than any other tanker firm during the years 1979–1982 through its operation of the Maersk Line and was responsible for 20 percent of crude deliveries over the period October 1979–February 1981. The firm furnished oil products as well and posted false destinations to help conceal the Maersk Line's involvement in South Africa's oil trade.[28] Especially noteworthy was an arrangement worked out in June 1979 by Helge Storch-Nielsen, a Dane serving as the honorary Peruvian consul in Cape Town, by which Sasol would provide 44 million barrels of Omani crude at a time when South Africa was trying to cope with the effects of the Iranian cutoff. One delivery per month was made during the last quarter of 1979, but the contract was then replaced by a deal in which John Deuss became the main conduit for Omani oil.[29]

In May 1984 leftist parties pressed for oil sanctions. The Danish Shipowners' Association pointed to the lack of a UN embargo against South Africa and maintained that economic sanctions would lead to the loss of Danish jobs. The Danish government responded with a policy opposing the use of Danish tankers in the South Africa oil trade (though not carrying penalties for violators). In September 1985 the A. P. Moller Company announced its withdrawal as a supplier; stronger government sanctions followed in June 1986 banning deliveries by Danish tankers and the sale of Danish North Sea crude.[30]

U.S. companies once played a secondary role in the provision of oil products, crude, and tanker transport (they are still active in exploration, refining, and marketing). Prior to the 1986 ban on both products and crude, the United States was a minor supplier. Phibro-Salomon furnished oil products until February 1985, and crude occasionally was sent. One case receiving attention involved the Panamanian-flag *Botany Triad*: The small tanker left Houston loaded with crude in November 1985 and headed for Durban. Off Port Elizabeth, it collided on December 6 with the Chinese cargo vessel *Lu Shan*, which was on its way from Shanghai to Libreville, Gabon; the Triad left an oil slick of 36 square miles when it lost 1,000 tons of its cargo.[31]

U.S. tankers delivered oil to South Africa during the years 1977–1981, after which the major oil companies discontinued using their own vessels. Involvement by U.S. companies since has been only sporadic. In 1983, for example, a ship owned by a U.S. company (but flying a Liberian flag) made three deliveries from Iran and the United Arab Emirates, while another vessel under time charter to a U.S. firm (but sublet to a different company) transported crude to South Africa from Brunei.[32] Overall, the United States has not contributed substantially to South Africa's acquisition of imported crude, though it aided considerably in the development of Pretoria's domestic oil sector.

MEN IN THE MIDDLE

Faced with greatly constricted imports after the Iranian revolution, South Africa turned to John Deuss and Marc Rich to overcome its most serious oil crisis. Deuss, a Dutchman who had sold Soviet oil until late 1977, quickly stepped in to provide some spot market cargoes acquired in Europe. He then worked out long-term contracts with Gulf states, first with Oman and then increasingly with Saudi Arabia. Operating through Transworld Oil, Lucina Ltd., and other companies, Deuss delivered about 6 million tons of Omani crude early in 1980 and approximately 12 million tons of Saudi oil in December of that year. This Saudi contract remained in force through March 1983, making Deuss the provider of 25 percent of South Africa's crude imports from 1979 until that time.[33] By far South Africa's biggest supplier, he earned considerable sums because the price of the Saudi crude provided was locked in—when oil prices began to decline, the South Africans tried (unsuccessfully) to cancel the contracts.[34]

Deuss's deliveries tapered off after 1983, but he still circumvented the oil sanctions and thus was the target of antiapartheid organizations. In January 1985 radical critics of his role firebombed his home in Berg en Dal, the Netherlands, and partially destroyed it.[35] Deuss remained adamant in defending his oil operations: "I am personally against apartheid. But I don't see the sense of depriving South Africa of oil. Sanctions don't work, as Rhodesia showed."[36] Gradually, however, Deuss disengaged. In 1985 he turned his attention to opportunities in the United States, where he purchased a refinery and 576 service stations in the Philadelphia area from Atlantic Richfield. He indicated in October 1987 he would stop supplying South Africa with crude because of its apartheid system,[37] but his future role became unclear in 1988 when it appeared he was selling his U.S. facilities to the Sun Oil Company.

Marc Rich, a U.S. oil broker, also helped South Africa at its time of greatest need. In April 1979 he negotiated an initial $1.5 billion contract to provide 50 million barrels of crude over one year; his continued deliveries through 1986 made him responsible for about 6 percent of South Africa's imports over that period.[38] Operating through various firms such as Minoil, Inc., Rich secured oil from many sources, though Brunei was a dominant supplier. By 1983 more than half of the oil Pretoria secured from Brunei was being furnished by Rich, and he controlled the flow from Brunei completely by 1985.[39]

While Rich was profiting from oil sales to South Africa, U.S. authorities were indicting Marc Rich and Company for evasion of taxes on oil income. His company submitted a guilty plea and agreed to pay $150 million in back taxes plus interest. Personal charges against Rich for making false statements were also pressed, but he escaped U.S. jurisdiction by finding refuge in Switzerland.[40]

Rich and Deuss became the subjects of an investigation in South Africa in the spring of 1984 after Progressive Federal Party opposition leader Frederick van Zyl Slabbert sent a dossier to Prime Minister Botha regarding oil trade practices. Van Zyl Slabbert alleged extensive mismanagement of oil procurement and mishandling of public money. More specifically, he claimed that Rich and Deuss were paid over the contract price, not counting funds expended through the Equalization Fund. Because five government or quasi-government officials also were cited in the dossier, van Zyl Slabbert called for two separate investigations: One, by the advocate-general, was to look into possible corruption; the other, by a select parliamentary committee, was to examine the use of public funds. Though Botha agreed to forward the dossier to the advocate-general, he ruled out the appointment of a special committee because he felt this action would be contrary to provisions of the Advocate-General Act.[41]

Piet van der Walt, the advocate-general, at first found "a prima facie case of enrichment at the expense of the state," but his investigation did not include interviews of Rich, Deuss, or some key officials of the Strategic Fuel Fund. On June 27 van der Walt reported that no such enrichment had taken place and that the amounts paid to Rich and Deuss did not exceed the sum stipulated in their contracts. There was thus no evidence of illegality and no prosecutions were warranted.[42]

MATTERS OF DISPUTE

Rumors and suspicions have circulated about other possible sources of South Africa's oil supplies. Marino Chiavelli is an intriguing character who moved to South Africa from Italy in April 1980 and applied for permanent residence after allegedly arranging a December 1979 meeting between Sasol's London manager and the director of the Saudi oil company, Petromin.[43] Chiavelli was granted permission to live in South Africa, but his residence application became an issue—he listed no criminal convictions—when it was later charged that he had been found guilty in Italy in cases ranging from check fraud to corruption to falsely using the title of "doctor." His background was reviewed in 1983, but the South African government decided not to revoke his permanent residence despite inaccuracies on his application[44] Other controversies soon arose, however, when it was revealed that Chiavelli paid only 176 rands in income tax one year and 75 cents the next despite reputedly being a billionaire.[45] There was also the unusual incident in which the police cleared him of suspicion that he had links to a contract murder syndicate.[46]

Chiavelli cut a dashing figure—he lived in a lavish mansion, hosted gala receptions, collected fine art, and became perhaps the fifth richest man in the world.[47] However, his alleged involvement in securing oil for South Africa was never completely clarified. In 1982 the chairman of the state oil exploration company (Soekor) said that Chiavelli had never supplied oil, but opposition member of parliament John Malcomess responded that this statement did not rule out Chiavelli's having acted as an agent for purchases.[48] In his 1984 report Advocate-General van der Walt indicated that Chiavelli had reached price agreement in negotiations with the Strategic Fuel Fund in 1979–1980, but that Chiavelli was unable to arrange the delivery and his discussions with the SFF ended in October 1980. Van der Walt concluded that Chiavelli had never sold crude to South Africa and had not purchased any oil on South Africa's behalf.[49] Yet there is room for speculation that Chiavelli may have played the role of middleman: For example, he allegedly facilitated a three-year 100 million-barrel deal with Saudi Arabia. In another case, ten bills of lading related to cargoes delivered in 1979–1980 purportedly named the same company that Chiavelli cited in a 1980 immigration document when he asserted he was its director.[50]

Then there is the strange affair of Greek-born South African Taki Xenopoulos, who was managing director of Fontana Holdings Ltd. In October 1983 his company sued Chiavelli in an effort to recover a $90 million commission it claimed was due on the 1981 three-year Saudi deal.[51] Xenopoulos was thus the apparent source of media information about the purported oil role played by Chiavelli, and Advocate-General van der Walt asserted that Xenopoulos also sent a file on Chiavelli to some members of the cabinet. Van der Walt maintained that it was in Xenopoulos's interest to spread such allegations because they buttressed his own legal case—he would be entitled to a commission if it could be proven that Chiavelli actually had oil delivered to South Africa. The advocate-general found otherwise. It was also reported that Xenopoulos may have furnished some of the materials included in the dossier on oil practices given to Prime Minister Botha by van Zyl Slabbert in April 1984.[52]

The Witwatersrand Local Division of the Supreme Court heard Xenopoulos's suit against Chiavelli, but information about the case is scarce because the Petroleum Products Act was used to justify extreme secrecy. The press was banned from even indicating when the court would meet, and the hearings were private. Just before a verdict was due, settlement was reached, but its terms have never been publicly revealed.

Another closed hearing was conducted in the same court beginning in March 1984 in a suit against the Strategic Fuel Fund, Sasol, and the honorary Peruvian consul, Helge Storch-Nielsen. Brought by a British lawyer named Maurice Sellier in collaboration with Middle East businessman Ezra Nonoo, South African Korean War air force hero Brigadier J.P.D. Blaauw, and Trade and Technology (Holding) Ltd., the case involved a June 1979 three-year contract to deliver Omani oil and its termination after only three months of deliveries. The precise involvement of Sellier is obscure, but he sued on the basis of anticipated commissions not received because the contract was replaced by one in which John Deuss provided Omani crude for South Africa. Sellier and his fellow litigants lost their legal fight, although the November 1984 verdict was not announced.[53] As later revealed in the minutes of a parliamentary session, the South African public lost no money in the case, although it could have been subjected to a judgment as high as 89 million rands because the SFF and Sasol were among the defendants.[54]

South Africa managed to secure oil with the covert aid of disparate countries and companies and was able to switch quickly to different suppliers and tanker fleets whenever the situation warranted. A plentiful quantity of crude continued to flow. The intriguing question that next needs attention is this: What techniques were used to circumvent sanctions and maintain a veil of secrecy over the clandestine oil trade?

9

CHICANERY ON THE HIGH SEAS
Techniques of Evasion

Oil sanctions against South Africa produce their inevitable counterpart: circumvention by devious and sometimes illegal means. South Africa, highly dependent on imported oil, cannot be absolutely ethical in the means of obtaining crude, and the veil of secrecy it throws over its transactions helps conceal a significant degree of duplicity, and even fraud. In its effort to acquire oil, South Africa is forced to seek suppliers outside normal channels; it has little choice but to deal with disreputable brokers and to overlook some shady methods of oil procurement and transport. In some instances South Africa itself becomes the defrauded party, but its gray market search for crude goes on.

CAT AND MOUSE

Numerous techniques are used to conceal deliveries to South Africa so that collaborators with Pretoria can avoid being blacklisted as accomplices of apartheid. Many methods parallel those employed in other commercial fields: An individual may create a trail of confusion by operating several companies simultaneously, or cargoes and vessels may be sold frequently in order to frustrate efforts to identify the owners. The charter system complicates matters—as owners of vessels are reluctant to reveal the names of charterers, and arrangements are sometimes made in which the charterer is actually working secretly on behalf of a tanker's owner. The role of charterers and management companies leaves shipowners with virtually no control of their vessels because the master takes his orders from the management company.

Payments in the oil netherworld are made through numbered bank accounts that are difficult to trace, and "post box" companies are frequently used to mask ownership. These companies are often registered in Liberia or Panama where their assets and officers escape legal scrutiny. Many are just shells used for legal convenience, making it difficult for investigators to follow the paper trail. Changing the names of tankers after they have visited South Africa adds to the confusion, as does the sale or scrapping of vessels after they discharge their crude in Durban or Cape Town. For example, the *Dirch Maersk* made its last voyage to South Africa in April 1980 and was scrapped the following month[1]—a frequent pattern, because older ships tend to be engaged in the South African oil trade so that newer vessels avoid the risk of being blacklisted at various ports in retaliation for their involvement.

As sanctions have come to be applied more comprehensively, it is rare for any tanker to list South Africa as its intended destination. This gives rise to spurious declarations that the cargo is not bound for South Africa (written disclaimers are required by several countries), and many posted destinations given when leaving port are patently false. In November 1982 the *Mobil Weser* was loaded in Rotterdam and ostensibly was bound for Bonaire as it headed to Durban. The *Cherry Vesta* claimed its destination as Japan when departing from Indonesia in February 1979, and the sister ships *Karoline Maersk* and *Karen Maersk* indicated when leaving Gulf ports in April and September 1980 respectively that they were going to Singapore and Italy. The *Cast Fulmar* sailed from Rotterdam in April 1981 with a destination of "America," but it changed its posting to Dakar, Senegal. Dakar's maximum cargo handling capacity was 35,000 tons; the *Cast Fulmar* carried over 108,000 tons of crude.[2]

Forged certificates facilitate the circumvention of sanctions. Documents indicated that the *Cast Puffin* had discharged its crude at Le Verdon, France, in September 1980 rather than in South Africa, but letters to Shirebu from French port and customs officials asserted that the tanker had never docked at Le Verdon. The *Ardmore* carried oil from South Africa to Tanzania in October 1983, but its documents claimed the oil originated in Singapore. In the *Salem* affair, (see Chapter 10) the bill of lading received upon loading at Mina al-Ahmadi, Kuwait, was altered in terms of cargo ownership.[3] Another certificate scam is to keep a small amount of oil on board when discharging in South Africa, take it to a refinery in a different country, and then receive false certification that a larger quantity has been delivered for refining.

Tankers attempt to disguise themselves when headed for South Africa. The *Salem* temporarily repainted its hull with the name *Lema*, while the *Sangstad* in July 1980 painted out its name (as did the *Norse King* a few months earlier) and took down its flag. On one of its voyages to Durban in 1979, the *Havdrott* lowered tarpaulins to conceal its name on both the hull and lifeboats.[4] Frequently, tankers cut off radio contact or use a coded designation, as was done aboard the *Karen Maersk* in September 1980.[5] Double logbooks are often kept, and occasionally a misleading course is taken to hide a delivery to South Africa. This may have been the case in the voyage of the *Jeppeson Maersk*, which transported oil products from St. John, New Brunswick, to Durban in May 1980: After departing South Africa, it entered the Mediterranean but picked up no cargo there. It was possibly feigning a Mediterranean delivery, and it then headed for the Bahamas.[6] The most drastic technique is scuttling—intentionally sinking tankers after they have left most of the cargo behind in South Africa and using remaining crude to create a slick that conveys the impression the tanker was fully laden when it was purportedly the victim of an accidental sinking. In addition, an insurance claim on the hull may even be filed.

Various schemes are implemented to hide deliveries and make it difficult to trace the origin of the oil. In December 1983 the *Manhattan Viscount* was loaded at Qatar; after bunkering at Ras Tanura, Saudi Arabia, the vessel transshipped its crude at the end of January to the *Thorsholm*, which proceeded to Durban.[7] A further refinement on transshipping is to mix crudes within the same cargo, a technique used in April 1980 when the *Karoline Maersk* partially loaded in the Gulf and then added 130,000 tons received from the *Havdrott*. A similar but more complex transfer involved the *Karen Maersk*, which in August 1980 secured oil from a Chinese tanker, pumped in more at an Iranian port, and then sailed to Bahrain where crude from the *Fleutje* and the *Havdrott* was mixed with the cargo.[8] What is especially interesting is that these maneuvers usually involved different shipping companies and several flags of registry.

"Multiporting," in which a tanker loads partial cargoes at several ports, is another common technique, as is "double loading," in which a ship discharges in South Africa but then reloads in Nigeria before delivering oil to Western Europe. (Double loading may have occurred when the *Norse Falcon* in October-November 1981 sailed to Nigeria after discharging its cargo in South Africa.[9]) "Topping" features pumping more than the registered quantity of crude into a tanker and storing the surplus until a sufficient supply is available to constitute a cargo bound for South Africa. Topping can also be accomplished by port personnel inflating figures when loading crude into a tanker; the port retains the difference, stores it, and later includes it in a cargo for South Africa. Finally, there have been reports of ships intentionally foundering off the African coast, with their oil then being transshipped and brought to South Africa for sale.[10] But evidence is insufficient to support these contentions because good reasons seem to have existed for transshipping oil in these cases: In April 1981 the *Energy Endurance* did transfer its cargo to the *Regina*, but it had indeed suffered serious underwater damage (and the oil probably was not sold to the South Africans anyway). The tanker *Esso Nederland* experienced a fire and was towed to Durban in June 1981; however, it was traveling in ballast on its way from Aruba to pick up a new cargo in the Gulf.[11]

To maintain secrecy, crew members of tankers plying the South Africa trade are pressured and bribed; and Oliver Tambo, president of the ANC, has claimed that their lives are jeopardized by radio silence, the filing of false information, and scuttling.[12] There is also a tendency toward fraud because tighter sanctions produce increasingly sophisticated—and often illegal—subterfuges.[13] In August 1979 international oil companies operating in South Africa sent a letter to the secretary for industries warning about the possibility of fraud because they discerned "an abnormal risk directly related to the current *modus operandi* of our crude importers." They pointed out that protective insurance could be obtained, but this would necessitate revelations about South African oil procedures. To further national interests and preserve secrecy, they preferred to assume the risk of fraud.[14] Ironically, the South African government itself fell victim to fraud in the *Salem* case when it opted to pay $30.5 million to Shell quietly rather than have it become known in open court that a stolen Shell cargo had been sold to the Strategic Fuel Fund. Corrupt practices may also spread to South African officials. Though the advocate-general concluded in his 1984 report that such allegations were groundless, he contradicted himself by affirming that there was reason to suspect that some enrichment had come at the expense of the state in the *Salem* affair.[15]

On one occasion, the master of a tanker refused to deliver crude to Durban, and legal wrangling resulted. The Norwegian firm of Fearnley and Eger time-chartered several vessels to transport crude to South Africa. Among the ships was the *Manhattan Viscount*—owned by Japan's Sanko, the largest tanker fleet in the world—which in September 1983 carried a cargo from Brunei to South Africa without incident. But complications developed on a January 1984 voyage when the master listed Singapore as his destination upon leaving Qatar, was later told to go to Rio de Janeiro, and was then ordered to head for Durban. Sanko objected to this directive from Fearnley and Eger, which managed the vessel, although it was leased to the Intercontinental Transportation Corporation Ltd. of Liberia. The master refused to discharge in Durban (though the oil got there anyway by being transshipped to the *Thorsholm*). Lawsuits pitting Fearnley and Eger against Intercontinental in one case and against Sanko in another were then filed in New York and Oslo respectively.[16]

It would be easy to police tanker visits to South Africa because nearly all are made to Durban or Cape Town, but in the absence of any international sea patrol, monitoring the crude oil trade is a rather difficult undertaking. Because South Africa refuses to list tanker calls, organizations like Shirebu must extrapolate information from *Lloyd's List, Lloyd's Voyage Records,* and *Lloyd's Register of Shipping* in order to determine the tankers and shipping companies involved and the flags of registration. Often clues are provided when a tanker recorded as leaving the Gulf returns approximately a month later without any indication that it has discharged crude at any port. This occurred on five occasions in 1983–1984 with tankers owned by the Norwegian company Thor Dahl.[17] But because many tankers stop in South Africa for bunkers, food, water, or new crews, determining which vessels have actually discharged crude there requires care: The best rule of thumb is to suspect those tankers whose last call and subsequent call are at ports of oil-producing countries.[18] Oil products present an additional problem because they are often delivered on combined carriers capable of transporting either liquid fuel or solid materials; therefore, that a combined carrier has delivered a cargo to South Africa is not proof that oil products were involved.[19]

Tips from crew members are an important source of information. An amusing—though inadvertent—revelation came when a seaman who wrote to a Norwegian radio show requesting a song indicated that his ship was on its way from South Africa to Singapore. His letter, read on the air in September 1979, indirectly provided evidence that the *Staland* had discharged in Cape Town after departing Iran.[20] Ships calling in South Africa generally leave a trail—they require food and bunkers, personnel may receive medical care, and officers may use transportation from ship to shore. Mail is even received while tankers are waiting to discharge at the Durban buoy, as occurred in a helicopter drop to the *Dagmar Maersk*.[21] It is rare that an investigator is permitted to delve into such matters at South African ports, but Scotland Yard detective Peter Griggs managed in February 1980 to gather irrefutable documentary evidence that the supertanker *Salem* had called at Durban.[22]

The ill-fated voyage of the *Albahaa B*. represents a tragic case in which disclosure was made about an oil delivery to Durban. It left Dubai in March 1980 claiming a destination of Singapore and then discharged in South Africa. The *Albahaa B*. then left in ballast for Oman, but was racked by explosions precipitated by poor procedures used in cleaning a tank. On April 3 it sank off Tanzania. Five crewmen were killed by the explosions, and another perished while abandoning ship. Publicity surrounding the sinking forced the ship's management to announce that it had in fact delivered crude to South Africa.[23]

WAY STATIONS

Oil is often sent to South Africa indirectly from tank farms. The Shell facilities at Pulo Bukom in Singapore have played an important role, as have the former Shell storage depot at Curaçao in the Netherlands Antilles (turned over to the government in 1985 after financial problems arose) and the one owned by Exxon at nearby Aruba. Bilbao in Spain serves a similar function in Europe, but the most important storage site there at which cargoes can be switched to different tankers is the Europoort at Rotterdam. During the years 1979–1982, five million tons destined for South Africa may have been transshipped there, even though storage adds about 5 percent to the cost of a voyage because of fees, port duties, and extra sailing time.[24] Rotterdam storage helps camouflage the sale of North Sea crude to South Africa: In September 1980 British oil was sent there; in October it was pumped aboard the *Robert Maersk* along with other crude and delivered to South Africa. In April 1981 Norwegian oil was dispatched to Rotterdam; in less than a day, some of it was loaded on the *Cast Fulmar*, mixed with Middle Eastern crude, and forwarded to Durban.[25]

In July 1979 a serious collision of tankers led to revelations about the transshipment of oil in the Netherlands Antilles. One ship involved was the *Atlantic Express*, which was transporting Gulf crude to Texas. The other was the *Aegean Captain*, which loaded Iranian crude at Curaçao and then added a smaller partial cargo at Bonaire. Allegedly chartered by Marc Rich, the *Aegean Captain* left Bonaire on July 17 with a stated destination of Tenerife in the Canary Islands. The next day a change was made to Singapore, but it appears that the vessel was intending to deliver its crude to South Africa.[26] The two ships collided July 19 twenty miles east of Tobago during a storm. The *Atlantic Express* sank with the loss of 26 lives; the *Aegean Captain*, its missing electrician presumed dead, was towed to Curaçao and scrapped—only 7 percent of its crude drained into the sea. The *Aegean Captain* flew a Liberian flag, the *Atlantic Express* a Greek one. Liberia therefore conducted an extensive investigation into the case, and a Greek court later sentenced senior officers of both vessels to prison terms for negligence, manslaughter, and violation of safety regulations.[27]

South Africa, always on the lookout for transshipment points, was presented briefly with an opportunity to develop a new logistics arrangement in Haiti in 1981. On the basis of the San Jose agreement of that year, nonproducing regional states such as Haiti were to be given a preferential price (a 30 percent discount) on Mexican oil. Two Haitian businessmen (purportedly one of them was Jean-Claude Duvalier's father-in-law) purchased Mexican crude as if it were intended for the Haitian market, but then shipped it to Curaçao for refining and sale to South Africa at a great profit. But this scheme soon ran into difficulties: Mexico, furious that its oil was being rerouted in this manner, billed the Haitians at the full market value rather than at the discount rate, and a bank on the Cayman Islands refused to handle the payments being made by South Africa.[28]

More recently South Africa has been turning to the nearby Seychelles. Relations between the countries have been improving since a low point in 1981 when South Africans were apprehended by the Seychelles in a failed coup attempt, and oil seems to be reaching South Africa through the Seychelles. One conduit has allegedly been provided by Craig Williamson, a former employee of the South African security services; he became the manager of the South African branch of a holding company based in the Seychelles, and he has links to the national oil company there.[29] Another connection may have been set up by Marc Rich, who purportedly went to the Seychelles in 1984 to discuss the purchase and resale of Mexican crude. He was working with Francesco Pazienza, a former member of Italy's military intelligence, who faced legal problems at home for alleged involvement in the financial collapse of a bank and in the bombing of Bologna's rail station.[30]

THE CARIBBEAN CAPER

Sydney Burnett-Alleyne was an entrepreneur from Barbados who founded the Alleyne Mercantile Bank. It is alleged that the bank never operated in a normal commercial fashion and that it may have served as a cover for the movement of money from Portuguese territories in Africa. Alleyne did have close ties to some Portuguese in Angola and Mozambique and to the breakaway government of Rhodesia.[31] In July 1985 he joined with Prime Minister Patrick John and Attorney General Leo Austin of not-yet-independent Dominica to establish the Dominican Development Corporation. Alleyne was expected to attract funds for the construction of an oil refinery and an international airport because there was a basic plan to attract foreign investment through the offering of tax concessions.[32]

Alleyne's bank in Barbados was closed when he failed to pay certain debts—such as rent— and he was soon imprisoned for a year in Martinique for plotting an invasion of Barbados. Alleyne resurfaced in London and went to work in August 1978 for a financial concern linked to Afro-Portuguese and South African interests. He then began to set up a complex scenario in which he would assist the South Africans in the development of oil facilities in Dominica, and they would arm him for an invasion of Barbados. He therefore reestablished contact with John and Austin and invited them to London, where they were introduced to officials from the South African embassy. Alleyne also formed a liaison with John Banks, a mercenary. In November Dominica became independent, and John signed a December agreement with Alleyne on Dominican economic development.[33]

Britain's Special Branch gained knowledge about the Barbados invasion plan and questioned Alleyne on December 11. Five days later Prime Minister Tom Adams of Barbados revealed the plot, code-named "Hilton." Apparently Alleyne hoped to create a federation between Barbados and Dominica, and John Banks has confirmed that South Africa was paying for the commando action.[34]

The military aspect of Alleyne's escapade was abandoned because of the publicity, and the economic arrangements with Dominica also were exposed by a newspaper in Barbados that same month. Nevertheless, efforts went forward on the commercial front as a representative of Alleyne telexed John on January 3, 1979, asserting that money had been received and the South Africans were prepared to finance an oil refinery and a tanker terminal.[35] On January 18 Alleyne wrote to Sasol's London manager, G. L.

Williams, to confirm an "offer to buy for and sell crude to SASOL with the approval of the government of the Commonwealth of Dominica."[36]

On February 6 Leo Austin wrote on Dominican government stationary to G. J. Coetzee, commercial minister at South Africa's embassy in London. He discussed the stockpiling, resale and refining of oil and indicated that the deal would go into effect on February 9.[37] That very day John and Austin secretly established a free port authority with complete jurisdiction over 45 square miles for 99 years at the rock-bottom rent of $100 per year. Don Pierson of Texas was to control this operation, and there were to be no taxes or duties within the free port. Pierson's deal was with Prime Minister John, and his son admitted that there was a plan to open gambling casinos within the zone.[38]

Evidence indicates that the South Africans were to use the free port as a minikingdom where they could carry out their oil trade under little scrutiny. Included in the construction projects were to be an oil refinery, storage tanks, a petrochemical complex, and a deep-water harbor for tankers. The South Africans were also committed to building an international airport and hotels.[39] The main purpose was to establish a way station for oil deliveries, and there was speculation that a company in Cyprus with ties to Dominica and Alleyne was in a position to provide 7 million tons of Saudi crude.[40] It was also conjectured that the South Africans were to finance the scheme through diamond sales.[41]

The South African government denied any involvement in the Dominican affair, but confirmed that private South African companies were participating.[42] Alleyne, when queried about the South African role, admitted that one of its companies was to construct the airport, and he acknowledged he was planning to sell refined oil to South Africa for hard currency. On the issue of whether Sasol held an interest in the planned refinery, Alleyne responded: "Well I wonder if I should answer that one? But we welcome all institutions of Sasol's capability and integrity."[43]

The South African connection rocked Dominica, especially after a British television program pointed out the involvement of senior government ministers. As a Dominican newspaper made new disclosures, trade unions organized a protest demonstration of about 15,000 people on May 29 in the capital, Roseau. Violence resulted, and one was killed and nine wounded. The demonstrators were opposing government efforts to silence critics of the free port plan by imposing tougher antistrike legislation and libel laws, but the harsh response of May 29 failed to deter the protesters—a general strike already in progress was continued with renewed vigor. In the turmoil Minister of Agriculture Oliver Seraphine quit the cabinet and verified that negotiations had taken place with South Africa. Prime Minister John tried to stem the tide by withdrawing his strike and libel legislation and by firing Austin, but he was unsuccessful and was forced to resign on June 16. Seraphine replaced him as prime minister and promised to review government contracts in the affair. The deal then collapsed because of adverse publicity. In essence the circumvention of oil sanctions had let to a political crisis in a tiny Caribbean state, and the removal from power of Patrick John stemmed directly from his role in facilitating South Africa's economic designs.[44]

10

THE CASE OF
THE SUPERTANKER *SALEM*
The Greatest Fraud
in Maritime History

The most dramatic example of maritime fraud was the *Salem* case.[1] The perpetrators chose to deal with South Africa knowing that few probing questions would be asked and that secrecy laws would help cover up their schemes. For its part, South Africa failed to protect itself against unscrupulous operators and even displayed some recklessness in collaborating with them. Little did Pretoria know that as it acquired much-needed crude, it was itself becoming an unintended victim of conspiracy.

FRAUD IN A NUTSHELL

The *Salem* was a Liberian-registered tanker of almost 214,000 deadweight tons. Its delivery of oil to Durban in December 1979 at first seemed to be just another example of sanctions busting, but an extensive fraud soon uncovered made the *Salem* case the most complex and fascinating maritime scandal ever. At least 25 countries were touched by the case, setting off 13 separate investigations; legal proceedings eventually were brought in the United States, Greece, the Netherlands, Britain and Liberia.

After the cutoff of Iranian oil, South Africa turned to the spot market and started to negotiate long-term contracts. Needing oil in a hurry, it was prepared to deal with almost any entrepreneur willing to furnish supplies. Into this scene of fuel scarcity and lax security stepped Fred Soudan, Anton Reidel, and Nikolaos Mitakis (from the United States, the Netherlands, and Greece respectively), in quest of their first oil delivery contract. They managed to secure one with the Strategic Fuel Fund, received financing from a South African bank to purchase a tanker, bought a shipping company, and acquired insurance. The fraud was already in motion as three men lacking experience in the oil trade were about to set up their big payday.

The perpetrators had the ship and the contract, but they had no oil to deliver. They therefore arranged to hijack a cargo by placing their own officers and crew on the *Salem* and then chartering it to a legitimate oil company seeking transport. Pontoil turned out to be the unsuspecting party; more than 196,000 tons of its crude were loaded on the *Salem* in Kuwait. By a quirk of fate, this cargo was sold to Shell four days later—after the vessel was at sea—so Pontoil emerged with a hefty profit and did not become the victim of fraud.

Shell believed that the *Salem* was taking its cargo to France, but disguised as the *Lema*, it was diverted to Durban by its master, Dimitrios Georgoulis. Over 180,000 tons of crude were discharged (16,000 tons remained in the tanks), and the masterminds behind the scheme were paid for the oil, minus the amount advanced to purchase the ship. At that point, the South Africans had no idea they had bought stolen crude.

The *Salem* was filled with seawater so that it would appear fully laden, but because Shell would obviously have been alerted to the theft of its cargo if the tanker were to start pumping out water rather than crude upon arrival in France, the ship was scuttled off the coast of Senegal at one of the deepest points in the Atlantic. The pretense maintained—that a full cargo of crude had been lost after mysterious explosions racked the tanker—was played out with the officers and crew leaving in lifeboats and being "rescued" by a passing British tanker. The perpetrators stood to make an additional $24 million on the *Salem's* hull insurance, but the unraveling of the fraud dictated against the filing of any claim.

Investigations into the *Salem* affair were undertaken by insurance interests and many concerned governments. Concerted action came where important financial interests were involved, whereas efforts were somewhat half-hearted in those jurisdictions lacking an economic stake. Eventually many of the leading perpetrators were convicted, including Soudan, Mitakis, and the officers and master of the *Salem*; Reidel's trial in the Netherlands was constantly delayed for procedural reasons. Soudan received the longest sentence after his conviction in Houston—35 years—but he escaped from prison and is still at large.

Hardly any money from the *Salem* fraud has been recovered. The major losers were South Africa, which paid Shell $30.5 million in an out-of-court settlement; Shell, which was still about $16 million short after receiving payments from the Strategic Fuel Fund and the cargo insurers; the cargo insurers themselves, who paid out approximately $10 million; and the hull insurance brokerage company, which apparently lost over $300,000 because it was obliged to compensate the underwriters for Soudan's policy even though he failed to pay most of his premiums. The total fraud therefore amounted to roughly $57 million, but the perpetrators garnered only $32 million because certain oil price differentials must be included in the calculations and the scuttled tanker represented a major lost asset.[2]

CIRCUMVENTING SANCTIONS

Fred Soudan was an insurance agent who decided to try his hand as a commodities broker. He therefore incorporated a firm in Houston, Texas, in February 1979 and set out to make a deal. Upon hearing that six 200,000-ton cargoes of crude were available for delivery to South Africa, Soudan eagerly sought the contract from the SFF in competition with numerous experienced brokers. Through newly established contacts in South Africa, he managed to get himself invited for negotiations in Sasolburg, although his dire financial situation forced him to borrow money from his cousin in order to pay for travel expenses.[3]

In October 1979 Soudan arrived in South Africa and made the most of his first brokerage opportunity. Meeting with SFF officials, the Lebanese-born Soudan claimed ties to Arab political figures as well as ownership of a refinery in Louisiana and a tanker. Soudan also professed ownership of an oil company that supposedly was constructing a new corporate headquarters in Houston, and he even invited his SFF hosts to its grand opening. Claiming he had a surplus of crude that his refinery could not handle he offered six cargoes of 200,000 tons each. The South Africans were prepared to go ahead with an initial delivery, but would not pay Soudan in advance as requested. An agreement was therefore worked out for 214,000 tons, plus or minus 10 percent, with delivery to be made in early February 1980. The date was soon advanced to December 1979 because the South Africans were anxious to secure the oil as soon as possible.[4]

Fred Soudan did not own an oil company, refinery, or a tanker, and he was not building a headquarters in Houston. Nor in fact did he have any oil to deliver. But the South Africans did not investigate his claims, despite his lack of background as a commodities broker. His request to receive payment through a Swiss bank was not unusual in the South African oil trade, and his plan outlined to the SFF regarding documentation was not of concern as long as the crude was supplied. Soudan had maintained that 10 percent of the cargo would be left aboard and shipped to Corpus Christi, Texas, where false papers would be furnished indicating that the entire cargo had been discharged there, thus hiding delivery to Durban. To carry out this scheme, Soudan said that he could not provide South Africa with the bill of lading because it was needed in the United States as part of his false-destination scheme. He maintained, however, that the SFF would be shown other proof that he actually owned the oil. South Africa's acceptance of this arrangement proved crucial to its becoming a defrauded party.[5]

Soudan claimed his tanker had a boiler problem and could not be used for the December delivery—he therefore asked the SFF to furnish a vessel. The SFF refused, but did agree to help with the extension of a letter of credit to purchase one. A South African bank then advanced the funds, with the stipulation that the $12.3 million be repaid off the top by the SFF before Soudan received payment for the discharged crude. Soudan then bought the *South Sun* and renamed it the *Salem*. He had parlayed his charm and business acumen into a major deal—he now had a ship, a delivery contract with the SFF, and tanks full of crude (which of course really belonged to Pontoil and later Shell).

While the *Salem* was at sea on its way to Durban, Soudan and Reidel went to South Africa to complete the paperwork for their oil transaction. At a meeting with SFF officials, Reidel had to prove that the crude in the anticipated delivery was his, so he presented a bill of lading. Because the original document was issued in Kuwait (when the tanker was loaded), mailed to Pontoil, and then sent on to Shell, Reidel could not possibly have had it with him in Sasolburg. The document he displayed therefore was apparently the master's copy. The South Africans were nonchalant about this matter even though they realized Reidel was not presenting the original bill of lading—in fact Reidel had scratched out Pontoil's name on the document as consignee of the cargo, which should have aroused suspicion. Nevertheless, the South Africans accepted a statement signed by Reidel and Soudan that the oil was the property of Reidel through his company, Beets Trading.[6]

The SFF remained in the dark about Shell's ownership of the *Salem*'s cargo when the tanker discharged at the single buoy mooring on December 28, 1979. Members of the crew of course were aware that the oil had been diverted to Durban, but their silence was bought through payoffs on the spot plus additional funds promised once the voyage was completed.[7] The South Africans had their crude—though under unusual circumstances—so they paid Beets Trading most of the money that very day through a Swiss account and furnished the remainder in two installments over the next two weeks (for a total of over $32 million) and the SFF reimbursed the bank that had provided the letter of credit to purchase the tanker.

THE LEGAL COURSE

Once Shell realized that its oil was in South Africa, it took legal action against the Strategic Fuel Fund. Negotiations led to an out-of-court settlement giving Shell $30.5 million from the publicly financed Equalization Fund. The cargo was worth at least $56 million, but Shell's loss was less than that suffered by the South Africans, who ended up paying twice for the same crude. The cabinet appears to have approved of this arrangement because the main objective was to avoid hearings at which South Africa's oil procurement methods would be disclosed. As expressed in an SFF report: "During that period, it was not in the country's interests that the particular vulnerability of the RSA in obtaining crude oil be exposed."[8]

South Africa investigated the *Salem* affair, although the desire to maintain oil secrecy prevented the handing down of any indictments. At first SFF director Dr. Dirk Mostert quickly examined the case, managing to interview Soudan, Reidel, and Mitakis, among others. Though Mostert quickly informed parliamentary representatives of opposition parties that fraud had occurred, he left open the possibility that Soudan would furnish South Africa with another shipment of crude.[9] It was not until September 1982, following the appointment of a more crusading energy minister, that a more thorough investigation was initiated. It was headed by P. C. Swanepoel, former chief director of the National Intelligence Service, who was then serving as an intelligence analyst for Prime Minister Botha. Swanepoel's efforts resulted in a report of more than 300 pages completed in November 1983.[10] It provides extensive details on Soudan's efforts to obtain the oil contract and a tanker, but it fails to address any issues that could prove embarrassing for South Africa. The SFF role was not dissected in regard to contacts with Soudan, the Shell settlement was not detailed, and the possibility that some South Africans may have acted illegally was not considered. The Swanepoel report has never been released publicly (although it was read into the transcript of a Greek trial), but the South African government did provide copies to Greece, the Netherlands, and the United States to assist in the prosecution of *Salem* defendants. A copy was also forwarded to the attorney-general of Cape Province so he could determine whether any indictments were warranted in South Africa.

South Africa otherwise did little to aid other countries with their investigations and prosecutions. The Swanepoel report was furnished on a confidential basis, but there was a reluctance to reveal more about oil operations than was absolutely necessary. When Peter Griggs, deputy chief superintendent of Scotland Yard's fraud squad, attempted to examine the case in South Africa less than a month after the sinking, the South Africans proved obstructive. Citing oil secrecy legislation, they refused to let him see important documents and sometimes insisted that statements by officials be made in Afrikaans rather than in English.[11]

South African cooperation was lacking also in the Dutch case against Reidel, perhaps in part because Swanepoel was snubbed in 1982 when the Rotterdam prosecutor's office denied him a copy of Reidel's file. Also relevant was the Klaas de Jonge affair. A Dutch citizen, de Jonge was suspected by the South Africans of furnishing arms and ammunition to the African National Congress. While in police custody in July 1985, he was seized by South African police, but then returned after diplomatic protests. The de Jonge affair therefore cast its shadow over Dutch legal proceedings in 1986–1987. The South Africans not only refused to permit the Dutch to obtain a statement from Swanepoel—more significantly, they blocked moves to secure SFF officials as witnesses to testify on the key point at Reidel's trial: Did he fraudulently alter the bill of lading?

At the U.S. trial of Fred Soudan, the role of South African witnesses became a matter of contention. Swanepoel, armed with a waiver from his government for possible violation of oil secrecy laws, did testify, but the three crucial SFF officials did not. Because they were not furnished with a similar waiver, they were willing to testify only if questions about oil procurement practices after the *Salem* affair would not be broached. Though the prosecution was prepared to abide by this guideline, the defense was not—it hoped to demonstrate a pattern of South Africa resorting to illegal methods and accepting misrepresentations made by oil suppliers. Because the three potential witnesses were not in the United States, they could not be subpoenaed, so their evidence was never presented to the court.

HUE AND CRY

Because public South African funds were used to compensate Shell, opposition member of parliament John Malcomess attempted to lift the official veil of secrecy thrown over the *Salem* case in the name of national security. Charging that the minister of mineral and energy affairs, Pietie du Plessis, owed an explanation to the taxpayers, Malcomess challenged the government's secrecy by discussing the issue in parliament in February 1983. His tactic gave the news media a way to circumvent censorship because parliamentary debates are printed in the legislative transcript and thus become public record. Du Plessis tried to ban publication anyway, but several newspapers defied his order, and the *Salem* case finally came to the attention of the South African citizenry.

Malcomess at least prodded the SFF to some action, but the memorandum on the SFF's role prepared and submitted to parliament in March 1983 was basically nonrevelatory. Malcomess in response presented details of British court judgments related to the *Salem* and also read aloud sections of the *The Piracy Business* by Barbara Conway. Clearly showing that public revelations about the case were extensive overseas, he argued that secrecy could not possibly protect South Africa's oil operations, but instead served to chill inquiries into the use of government funds and the behavior of South African officials. Malcomess called for an investigation by the advocate-general and the appointment of a select parliamentary committee; he was turned down on both counts because du Plessis believed oil secrecy was crucial in order to protect suppliers and shipowners.[12]

The van Zyl Slabbert dossier presented to the prime minister and the advocate-general dealt with allegations related to South African oil procurement, but not specifically with the *Salem* case (although the cover memorandum noted that the SFF lost a large sum in the *Salem* affair yet never complained to the police or sought an investigation[13]). Swanepoel lent his full-time assistance to preparation of the advocate-general's report, which barely touched upon the *Salem* except for the following observation: "From the information at my disposal concerning the *Salem* case it does appear as if certain persons might well have been improperly enriched at the expense of the State, but I did not deem it advisable or necessary to investigate the matter further since it is already in the hands of the Attorney General of the Cape with a view to possible criminal prosecutions."[14]

The attorney general of Cape Province looked into the *Salem* case and eventually announced in May 1985 (more than five years after the sinking) that no prosecutions were warranted because no crimes had been committed by South African citizens. Malcomess then charged that oil secrecy laws were concealing "at best negligence and at worst corruption" and that the political interests of the ruling National Party were the dominant factor in the case, not national security.[15] The *Salem* affair was thus laid to rest by the South Africans.

11

THE RHODESIAN PARALLEL
Comparisons with the South African Situation

Rhodesia—now Zimbabwe—was the target of protracted oil sanctions for fourteen years (1965–1979). Like South Africa, political dominance by a white minority and racist policies precipitated outside consternation, and sanctions were imposed by the Security Council of the United Nations. Not surprisingly, South Africa strongly assisted Rhodesia in circumventing them, and the same oil companies worked to undermine the embargoes against both countries. The lessons of the Rhodesian experience therefore have relevance for an evaluation of the current South African situation. Rhodesia clearly coped with the oil embargo in economic terms, but it simultaneously evolved politically toward black inclusion and, ultimately, independence under a black majority government. The impact of sanctions was thus complex, and their possible effect on political change must be subjected to careful scrutiny.

IMPOSING SANCTIONS

Britain and Rhodesia were heading toward a showdown in the fall of 1965 because Prime Minister Ian Smith appeared determined to declare independence under the rule of Rhodesia's white minority. In order to deter the breakaway colony, Prime Minister Harold Wilson journeyed to the Rhodesian capital, Salisbury (now Harare), and conferred with Smith on October 29. He warned that oil sanctions would be applied if there was a unilateral declaration of independence (UDI), but Smith did not succumb to the pressure, and UDI was affirmed on November 11.[1] Rhodesia's action was buttressed by advance assurances of an adequate oil supply should sanctions be imposed: On October 14 Portugal gave its support—crucial, because it controlled the import route through Mozambique. Shell promised to continue deliveries, which Smith later declared influenced his UDI decision. Rhodesia had less than one month's supply of gasoline and other oil products, but it stockpiled just before UDI to triple reserves by November 20. Oil was stored at Umtali (the location of Rhodesia's refinery) and at the Mozambican port of Beira (the terminal for the pipeline to Umtali), and a tanker was just then preparing to discharge at Beira.[2]

Britain's reaction was purely ad hoc because no sanctions plan had been worked out prior to UDI. Sanctions imposed November 16 did not include oil, and the UN Security Council then began to force the issue by voting on November 20 to apply voluntary oil sanctions. Two days later Iran joined the bandwagon by announcing an embargo, but it actually continued to deliver oil indirectly through South Africa.[3] On November 23 Wilson said that an oil embargo would be difficult to apply and that the matter required careful study.[4] Considerable pressure was then brought to bear on Britain by the Organization of African Unity, which convened a session of its Council of Ministers on December 5–6. An oil embargo was endorsed in order to bring down the Smith government, and Britain was given ten days to take action. When it failed to do so, and even permitted a BP tanker to discharge at Beira, thirteen African states broke diplomatic relations on December 16. This appears to have been the triggering event—Britain imposed oil sanctions the next day. Significantly, however, they did not cover subsidiaries of British companies registered in other countries.[5]

Britain, which had decided to use a combination of pressure and diplomacy in lieu of force, secured support on the sanctions from the United States and France.[6] In an effort to discourage tanker deliveries to Beira, naval vessels were stationed off the Mozambican coast, and Wilson declared at the January 1966 Commonwealth summit in Lagos, Nigeria, that his country would not oppose further UN oil sanctions. He also maintained that an oil embargo could produce the collapse of UDI—though fifteen years later he admitted there really had been no way to remove Smith and claimed that oil sanctions were not applied toward that end.[7]

At the end of March 1966 the Labour Party was reelected, putting it in a position to take stronger action against Rhodesia. It called for a coastal blockade of Mozambique so that oil could not be discharged into the Beira pipeline, but when British vessels intercepted the *Joanna V.* on April 4, London felt it could not legally prevent the ship from docking at Beira in the absence of a Security Council resolution. When the master maintained that the tanker was not planning to deliver a cargo, but had to proceed to Beira for provisions and bunkers, the British allowed it through the blockade. The Security Council then called upon Britain to enforce a sea blockade and to arrest the *Joanna V.* if it discharged at Beira. The British immediately turned back the *Manuela*, but took no action against the *Joanna V.*, because it did not deliver oil. Whether it had intended to leave its cargo at Beira is conjectural, although the British foreign secretary implied that the international attention focused on the tanker prevented it from supplying Rhodesia.[8] Britain's blockade lasted until June 1975, when Mozambique received its independence from Portugal and refused to serve as a pipeline transit route for oil bound for Rhodesia. Nevertheless, Britain did not actually intercept any tankers after 1972 because the Feruka refinery in Umtali was closed and there was little reason to use the Beira pipeline when other means of funneling oil into Rhodesia were available.[9]

On December 16, 1966, the Security Council included a mandatory oil embargo against Rhodesia in the selective sanctions that were approved, the first time the Security Council ever voted for mandatory measures. On May 29, 1969, further action was taken, with comprehensive economic sanctions put into effect. Despite UN efforts, Rhodesia still managed to secure all the oil it needed.

AIDING AND ABETTING

Rhodesia, which had relied on oil products refined in South Africa and Mozambique, opened its own Feruka refinery in 1965 before UDI and started to import crude. But the refinery's closure in January 1966 because of the British naval presence off Beira made it again necessary to import oil products. Rhodesia coped extremely well in the face of economic pressure even as its consumption of oil doubled during the years of sanctions (from 9,000 barrels per day to 18,000). It also had to face internal guerrilla warfare, which included attacks by black liberation forces on oil installations. In one raid near Salisbury, rockets destroyed twenty-two storage tanks and damaged an additional seven; South Africans helped to extinguish the flames.[10]

Rhodesia's response to the oil embargo was to establish a central state procurement agency (Genta) in February 1966, to stockpile, and to raise the gasoline price to be the highest in the world.[11] There was also gasoline and diesel fuel rationing, which was instituted in December 1965, eased in March 1966 and again the following year, abrogated in May 1971, and then reimposed in February 1974 because of the Arab embargo.[12] The government also mandated that oil companies operating in Rhodesia had to comply with state directives regarding the distribution of fuel. As described by Brian White: "As far as oil and petroleum products were concerned, emergency powers legislation was speedily passed, making local oil companies controlled industries. This meant that they were legally bound to continue the supply of oil and faced legal penalties if they attempted to implement a sanctions policy."[13]

A sidelight to the oil sanctions was their impact on neighboring Zambia. Prior to UDI, Zambia received oil products from the Feruka refinery, but Rhodesia then imposed an embargo on December 16, 1965. The Salisbury regime obviously needed to hoard its own fuel supply, although it probably hoped as well that its move would be viewed as preemptive countersanctions and therefore serve to deter Britain and the Security Council from taking effective action. Six days later Zambia had to resort to fuel rationing, a severe problem that prompted Britain, the United States, and Canada to airlift oil and have it trucked to landlocked Zambia from Tanzania. In 1967 a pipeline project linking Zambia to Dar es Salaam, Tanzania, was implemented and oil started flowing in August 1968.[14]

Oil companies operating in Rhodesia prior to UDI continued to furnish supplies afterward, controlling the same percentage share of the market as before (Shell/BP 52 percent, Caltex 20 percent, Mobil 22 percent, and Total 6 percent[15]). All of the oil for the Rhodesian armed forces was provided by a subsidiary of Shell.[16] These same five companies marketed oil in South Africa, and 4 percent of that country's imports were sent on to Rhodesia. Prime Minister Henrik Verwoerd told the South African parliament in January 1966: "If there are producers or traders who have oil or gasoline to sell, whether to this country or to the Portuguese, Basutoland, Rhodesia or Zambia, then it is their business and we do not interfere."[17]

South Africa assisted its white minority Rhodesian ally, as did Portuguese-ruled Mozambique. The blockade of Beira and closure of the pipeline to Umtali did not prevent tankers from calling at the capital, Lourenço Marques (now Maputo), and the Sonarep refinery was therefore able to produce 6,000 barrels per day in excess of Mozambique's needs. This was forwarded to Rhodesia by rail. South African rail deliveries also were routed through Mozambican territory until 1976.[18]

From 1968 until 1971, Shell/BP carried out switch deals with the French company Total—Total supplied Shell/BP markets in Rhodesia, and Shell/BP compensated Total within South Africa. In a sense, Britain was thus in a position to deny it was fueling Rhodesia. From 1971 until March 1976, Shell Mozambique was Rhodesia's prime source of supply, even though Mozambique was subjected to an Arab oil embargo in 1973–1974. After independent Mozambique's border with Rhodesia was closed by the Frelimo government, switch deals with Sasol made South Africa the main oil lifeline. Oil products from the Natref refinery were sent north by rail, with much of the crude for Rhodesia being furnished to Natref by Brunei. Shell and BP provided oil to Sasol from their Sapref refinery, where production was increased from 1977 to 1978 despite South Africa's economic slowdown. Though these switch deals with Sasol became particularly important in 1976, the chairman of BP confirmed they had begun in 1971.[19]

Britain was supposedly embargoing Rhodesia, but subsidiaries of its two largest oil companies played leading roles in sanctions busting. In July 1977 Commonwealth heads of government meeting in London established a sanctions working group. Its eleven members, including Britain, called upon South Africa to stop supplying Rhodesia—an ironic request, because the Shell/BP-Sasol switch deals were then operative.[20] On the home front, allegations of British complicity in fueling Rhodesia led to the appointment of a commission and the publication of an investigatory report in September 1978. It exposed the operations of Shell and BP and charged that sanctions violations had been covered up by British authorities. These revelations brought an end to the Sasol switch deals, but there later were assertions in the press (denied by both the British and U.S. governments) that an agreement had been worked out between Prime Minister James Callaghan and President Jimmy Carter at the January 1979 summit of Western leaders—they would condone oil deliveries to South Africa if Pretoria in turn would pressure Rhodesia to accept a black majority government.[21] In December 1979 Britain's attorney-general, Sir Michael Havers, announced that Shell and BP would not be prosecuted for their actions in Rhodesia. In May 1980, after sanctions had been terminated and Rhodesia had become Zimbabwe, Parliament approved an amnesty for sanctions violators (excluding the limited number already prosecuted) on the ground that the lifting of sanctions meant questionable practices applied to Rhodesia were no longer crimes.[22]

As negotiations led toward a political solution in Rhodesia (independence as Zimbabwe was affirmed in April 1980), Britain eased up on oil sanctions. Prime Minister Thatcher indicated in August 1979 they would be ended, and they were partially phased out prior to the December 12 decision to abrogate them completely. With sanctions terminated, South Africa continued to be the conduit for Rhodesia's oil because the Beira pipeline and Feruka refinery remained closed.

ASSESSMENT

Did sanctions force Rhodesia to acquiesce to a plan calling for a black majority government? Certainly there was some economic disruption—Rhodesia was compelled to import goods at a premium and export them at a discount. Foreign reserves therefore dwindled, which prevented businesses from planning ahead with confidence.[23] On the whole, however, sanctions did not have a serious impact on the quality of life, and the prime motivation for Smith's willingness to end UDI can probably be found elsewhere. The Zimbabwe African National Union (ZANU) and Zimbabwe African People's Union (ZAPU) had sapped the government's energy through protracted guerrilla struggle, and independent Mozambique had turned to supporting antigovernment nationalist movements. South African backing had also weakened as Pretoria moved toward "detente" with black states. An additional factor was diplomatic pressure emanating from Britain and the United States.

The oil embargo of Rhodesia failed to achieve its aim because Britain was less than rigorous in applying sanctions and because South Africa and Mozambique came to Rhodesia's assistance. Many South Africans argue that the Rhodesian case demonstrates the futility of sanctions, but this line of reasoning is somewhat simplistic if it leads to the inference that oil sanctions against South Africa can never be successful.[24] First, it must be realized that monitoring the embargo against Rhodesia was an extremely difficult undertaking (despite the Beira sea blockade) because oil flowed in through two neighboring states. South Africa cannot expect such regional support, and a naval flotilla could interdict deliveries to its major ports. Second, South Africa requires a much larger oil supply, and its advanced economy is highly dependent on external investment and technology. Sanctions may therefore have a more deleterious impact than they had in Rhodesia because economic interdependence breeds vulnerability.[25]

The Rhodesian oil embargo spurred similar efforts against South Africa, and officials in Pretoria undoubtedly realized they had to sustain the Smith government in order to blunt any reinforcement of oil sanctions against their country. Ironically, implementating measures against South Africa would really have been the most effective way to topple Smith because the oil route from the south was his regime's lifeline.

South Africa, thanks to pliant commercial interests and lax enforcement of sanctions, has managed as well as Rhodesia, and the political concessions it has made cannot be directly attributed to the oil factor. Like Salisbury before it, Pretoria has turned oil sanctions into a double-edged sword: Smith used his power over the Zambian oil supply, while Botha has followed suit with his dependent neighbors—and his government threatens further steps if necessary. Just as Rhodesia survived, so too should South Africa be able to weather oil sanctions if they continue to be applied at their current pace and level of intensity.

12

SANCTIONS IN
THE BALANCE

Sanctions have been compared to international law and total disarmament: surely meritorious but overly idealistic.[1] Full compliance is unlikely to occur, just as the lamb is still loath to lie down with the lion and swords have yet to be beaten into plowshares. However, sanctions do tend to produce a healthy moral fervor, and they can help speed the rehabilitation of an offending party if accompanied by a sufficient degree of force.

THE MOMENTUM OF HISTORY

Embargoes habitually fail to achieve their political aims because disunity usually prevails among those applying sanctions and considerable laxity characterizes their implementation.[2] Target states also adjust successfully to outside pressures, thereby alleviating economic disruption and reducing the prospects for induced internal change.[3] Oil sanctions may raise fuel prices and make South Africa more susceptible to fraud, but they are really a mild form of punishment rather than a lever effecting the deinstitutionalization of apartheid.[4]

Though sanctions may be deficient in economic and political terms, they create a moral climate that gradually enervates their target. Without being driven to economic desperation or political calamity, most South Africans nevertheless have come to accept the inevitability (although not the desirability) of black majority rule. They realize that historical trends dictate the eventual demise of apartheid and that their system (plus that of the administered territory of Namibia) represents the last bastion of white minority rule in Africa. Whereas many South Africans previously endorsed apartheid out of conviction, most adherents now do so out of fear. If indeed "laager" mentality is growing among white South Africans, it represents the defensive action of those who recognize that their days of preeminence are numbered.

Sanctions serve to demoralize through extended isolation in the diplomatic, athletic, and cultural spheres, which has led to growing white emigration and frantic piecemeal dismantling of the peripheral vestiges of apartheid. The most telling sign that prominent South Africans are already thinking the so-called unthinkable is that leading members of the business community—including the two most powerful industrial leaders, Gavin Relly and Tony Bloom—have entered into a dialogue with the African National Congress aimed at eventual peaceful accommodation.

TOWARD A SOLUTION

Gradual, incremental sanctions are ineffective because the target state is given too much time to plan countermeasures prior to each strengthening of the sanctions guidelines.[5] Furthermore, the absence of mandatory sanctions provides a convenient rationale for states and corporations that seek to profit through circumvention—they can argue they should not be forced to the moral sidelines while others are left free to pursue venal self-interest. Sanctions are thus largely unsuccessful when each party fears that others will profit to its detriment.[6]

Can sanctions—particularly oil sanctions—be tightened sufficiently to render them a critical weapon in the struggle against apartheid? The continuing oil glut, plus past experience indicates that attempts to control sales by producing states are rather fruitless. So too would be closer supervision of documentation—fraudulent papers abound—and mandating that tankers state their true destination before leaving port could never be effective because a vessel may quite legitimately change destination once at sea if its cargo is sold or the ship redirected to a different market.

Port blacklisting of tankers that have called at South Africa is also a frustrating task: Many vessels manage to visit South African ports secretly, and those whose identities are known may later disguise themselves through name changes. It is also questionable whether a tanker should be blacklisted after coming into the hands of new owners who played no role in deliveries to South Africa, which raises the question of how deliveries to South Africa can be catalogued. A South African publication claimed that Soviet and U.S. satellites could effectively perform this function, but even if they could, they would have great difficulty distinguishing between tankers that discharge oil and those that make port calls for bunkers or provisions.[7]

Flags of convenience represent another set of problems. Liberia, Panama, and other open-registry states are not inclined to monitor the actions of tankers flying their colors because their aim is to raise revenue by attracting as many ships as possible to their registry. (In fact, it is surprising that African states and the OAU have not singled out Liberia for more extensive criticism—its ships continue to deliver oil to South Africa.) Flags of convenience also serve as a refuge for shipowners who may be pressured by their own country of registry. If restrictions regarding South Africa become too severe, the general inclination is to "flag out" by switching registration to FOC states. This cannot be prevented because shipowners have the right to register wherever they wish, and flags of convenience should be valid as long as minimum safety and labor practices are maintained.

The United Nations has responded weakly: Mandatory Security Council sanctions have not been imposed—and even if they were, violators could not be fined or punished. The General Assembly's appointed commission (cumbersomely titled Intergovernmental Group to Monitor the Supply and Shipping of Oil and Petroleum Products to South Africa) not surprisingly called for a mandatory Security Council oil embargo, but displayed traditional UN reticence to blame oil-exporting states for fueling apartheid.[8] The UN has always been constrained by the "domestic jurisdiction clause" of its charter, which limits its ability to deal with apartheid on more than a symbolic level because establishing some form of international police authority would require approval (unlikely to be forthcoming) of the Pretoria government.[9] Decades of UN condemnation of apartheid have produced few practical results, and Margaret Doxey appears correct in concluding: "Given all these difficulties, one is skeptical about resort to an oil embargo as an international sanction unless the pressure on Western powers to vote for it at the United Nations against South Africa makes it too costly not to. In such a case, the objective might be directed less to the coercion or punishment of the target than to satisfying those most actively seeking sanctions."[10]

Can sanctions contribute appreciably toward the dismantling of apartheid? Certainly, if backed by force—the Beira naval blockade during Rhodesia's UDI provides a useful precedent. The simplest solution would be for the Security Council to authorize the stationing of naval vessels off the South African coast with instructions to turn away all tankers except those seeking bunkers, provisions, or repairs.[11] If a vessel permitted to proceed for these purposes were to leave South Africa with less oil aboard than when it arrived, it would be seized. In practice, few ships would go to South Africa for bunkers because an oil shortage there would make the continued provision of such fuel unfeasible.

This basic type of sanction could work, whereas more complex undertakings invariably produce strains and schisms among their adherents and leave room for circumvention. Assuming a considerable degree of success in stemming the flow of oil to South Africa, one must question toward what end is such economic pressure being applied. Rehabilitation, not punishment, should be the keynote, but it would be unrealistic to expect the South African government to surrender the reins of power to the ANC. It would also be foolhardy to link an oil embargo to specific reforms demanded in the apartheid system, given the certainty of great disagreement among Security Council members as to the exact parameters. Here again the simplest solution is called for: The oil blockade should be linked to the demand that all South Africans be granted equal political rights and that a national election be held within six months. Let the South Africans decide their own future—which unquestionably will be based on overwhelming black representation and the dismemberment of apartheid.

South Africa can attempt to combat a blockade through a strategy of resource denial, but the potential of this weapon has already been reduced by preemptive Western stockpiling. More serious is the fuel problem that will be faced by Botswana, Lesotho, and Swaziland, which import their oil from South Africa; a plan to supply them (in Lesotho's case, by air) must be worked out prior to the imposition of a blockade. Finally, though an oil shortage in South Africa is bound to depress the economy and thus the standard of living, other means of ending apartheid would probably prove worse because they could precipitate civil conflict and the flight of white capital. If blacks are to acquire political rights and power, short-term economic sacrifice may be required.

A combination of sanctions and a naval blockade should be the most effective approach on the oil front. Sanctions alone have proven to be a weak instrument, but their importance should not be underestimated because they serve symbolically to rally moral support and increase the pressure for change.[12] Unless the groundwork is laid by the implementation of sanctions, a meaningful naval blockade would probably not even be considered. A psychological atmosphere is now being created in which South Africa is the focus of international opprobrium: Universities are selling their portfolio shares in corporations investing there, and divestment by major companies has become commonplace. In the United States, Democratic presidential candidate Michael Dukakis even labeled South Africa a "terrorist state." Though oil sanctions have basically failed to alter apartheid in any significant way, they have nevertheless played a meaningful role: Costs to South Africa have been raised, and sanctions have contributed to a broader international consensus directed against Pretoria's racially biased system.

NOTES

CHAPTER 1

1. *Guardian*, June 3 and 4, 1980; *Daily Telegraph* (Britain), June 4, 1980; and Martin Bailey, "Sasol: Financing of South Africa's Oil-from-Coal Programme" (New York: Centre Against Apartheid, February 1981), p. 6.

2. "Fuelling Apartheid: Shell and the Military" (London: Christian Concern for Southern Africa, 1984), p. 5; *Sun* (South Africa), June 4, 1982; *Rand Daily Mail* (South Africa), July 1, 1982; *Times*, November 9, 1982, and October 11, 1983; *Guardian*, March 12, 1984, and May 19, 1984; *International Herald Tribune*, May 15, 1984; *Financial Times*, November 29, 1985; *New York Times*, June 23, 1986; *Newsletter on the Oil Embargo Against South Africa*, no. 5 (September 1986): 8; *Africa Confidential* 24 no. 7 (March 30, 1983): 1; Johannesburg Domestic Service in English, March 11, 1984 (FBIS, MEA, 5, no. 049, March 12, 1984:U7), and May 14, 1984: (ibid., no. 094, May 14, 1984:U3); Harare Domestic Service in English, May 14, 1984: (ibid., no. 095, May 15, 1984:U5); and Umtata Capital Radio in English, March 11, 1984 (ibid., no. 049, March 12, 1984:U7), and March 13, 1984, (ibid., no. 050, March 13, 1984:U7).

3. "Fuelling Apartheid," p. 1.

4. For details, see "Shell Out of Namibia and South Africa" (London: Anti-Apartheid Movement, August 1980), pp. 10–11.

5. African National Congress, "Fuelling Apartheid," *Development and Social Progress*, no. 16 (July-September 1981): 113. See also African National Congress and South West Africa People's Organization, "Call for an Oil Embargo" (London, March 1985).

CHAPTER 2

1. See Martin Bailey and Bernard Rivers, "Oil Sanctions Against South Africa," (New York: Centre Against Apartheid, June 1978, p.75); and "Oil Sanctions and Southern Africa," (London: Commonwealth Secretariat, April 1974) p. 42.

2. Bailey and Rivers, "Oil Sanctions and Southern Africa," p. 44. In return, ANC Secretary-General Alfred Nzo thanked the Algiers summit for imposing an embargo against South Africa. See "Oil Embargo Against South Africa," *Sechaba* 8, no.5 (May 1974): 11.

3. OAPEC resolution, Kuwait, May 6, 1981.

4. States have differing rules regarding this matter in terms of the time lapse permitted between visits to South Africa and OPEC ports. For example, one state could blacklist ships that called at South African ports within the past month, while another state could stipulate three months.

5. Michael Tanzer, Terisa Turner, Bernard Rivers, and Martin Bailey, "Oil—A Weapon Against Apartheid," International Seminar on an Oil Embargo Against South Africa, Amsterdam, March 1980, pp. 42–46; and Michael Tanzer, Terisa Turner, Jennifer Davis, and Sybil Wong, "Toward an Effective Oil Embargo of South Africa," *Monthly Review* 32, no.7 (1980): 60.

6. *Secret Oil Deliveries to South Africa, 1981–1982* (Amsterdam: Shipping Research Bureau, June 1984), p. 46; Sanctions Working Group, "Means for the Implementation of an Effective Oil Embargo Against South Africa" (New York: Centre Against Apartheid, March 1981), p. 15; and Terisa Turner, "Trade Union Action to Stop Oil to South Africa" (Port Harcourt, Nigeria: University of Port Harcourt, 1985), p. 16.

7. *Lloyd's List*, November 29, 1982, p. 1.

8. Statement by Maritime Unions Against Apartheid, Conference of Maritime Trade Unions, London, October 30–31, 1985, p. 2.

9. *Lloyd's List*, October 31, 1985, p. 1; and statement by Jim Slater, Conference of Maritime Trade Unions, p. 3.

10. *South Africa's Lifeline: Violations of the Oil Embargo, 1983–84* (Amsterdam: Shipping Research Bureau, September 1986), p. 46.

11. "International Sanctions Against Apartheid South Africa" (New York: Centre Against Apartheid, May 1980), pp. 19–21; and Stanley Uys, "Prospects for an Oil Boycott," *Africa Report* 25, no. 5 (September-October 1980): 17.

12. Holland Committee on Southern Africa and Working Group Kairos, "The Dutch Campaign for an Oil Embargo Against South Africa, and the Withdrawal of Shell from South Africa," International Seminar on an Oil Embargo Against South Africa, Amsterdam, March 1980, pp. 2-4.

13. Bailey and Rivers, "Oil Sanctions Against South Africa."

14. United Nations General Assembly and Security Council, "Report on the Intergovernmental Group to Monitor the Supply and Shipping of Petroleum Products to South Africa," A/42/45 S/19251 (November 5, 1987).

15. See Articles 39 and 41 of the United Nations Charter.

16. Security Council Resolution 569, July 26, 1985, in *Resolutions and Decisions of the Security Council 1985* (New York: United Nations, 1986), pp. 8–9.

17. James Barber, "Economic Sanctions as a Policy Instrument," *International Affairs* 55, no. 3 (July 1979): 376.

18. See C. Lloyd Brown-John, *Multilateral Sanctions in International Law: A Comparative Analysis* (New York: Praeger, 1975), p. 1. A study of U.S. sanctions against Cuba and the Dominican Republic concludes that economic sanctions are generally not effective unless accompanied by diplomatic and military pressure. See Anna Schreiber, "Economic Coercion as an Instrument of Foreign Policy," *World Politics* 25, no. 3 (April 1973): 413.

19. Gary Hufbauer and Jeffrey Schott, *Economic Sanctions in Support of Foreign Policy Goals* (Washington: Institute for International Economics, October 1983), p. 83.

20. See James Barber and Michael Spicer, "Sanctions Against South Africa—Options for the West," *International Affairs* 55, no. 3 (July 1979): 389.

21. Peter Wallenstein, "Characteristics of Economic Sanctions," *Journal of Peace Research* 5 (1968): 261.

22. See Hufbauer and Schott, "Economic Sanctions in Support of Goals," pp. 78–79

23. Peter Calvocoressi, "The Politics of Sanctions: The League and the United Nations," in Ronald Segal, ed., *Sanctions Against South Africa* (Baltimore: Penguin, 1964), p. 52.

24. Gary Hufbauer and Jeffrey Schott, *Economic Sanctions Reconsidered* (Washington: Institute for International Economics, 1985), p. 10.

25. See Richard Stuart Olson, "Economic Coercion in World Politics: With a Focus on North-South Relations," *World Politics* 31 no. 4 (July 1979): 493.

26. David Baldwin, "The Power of Positive Sanctions," *World Politics* 24 no. 1 (October 1971): 25–26.

27. Johan Galtung, "On the Effects of International Economic Sanctions," *World Politics* 20, no. 3 (April 1967): 380-81.

28. See Barber, "Economic Sanctions as a Policy Instrument,"pp. 370–71.

29. Galtung, "On the Effects of Sanctions," pp. 411–412.

30. Fredrik Hoffmann, "The Functions of Economic Sanctions: A Comparative Analysis," *Journal of Peace Research* 4 (1967): 155.

31. M. S. Daoudi and M. S. Dajani, *Economic Sanctions: Ideals and Experience* (London: Routledge and Kegan Paul, 1983), p. 162.

CHAPTER 3

1. Richard Moorson, *The Scope for Sanctions* (London: Catholic Institute for International Relations, 1986), p. 39.

2. For earlier estimates, see Martin Bailey, "The Impact on South Africa of the Cut-off of Iranian Oil" (New York: Centre Against Apartheid, July 1979), p. 1; Martin Bailey and Bernard Rivers, "Oil Sanctions Against South Africa" (New York: Centre Against Apartheid, June 1978), p. 5; "Oil Sanctions Against South Africa" (New York: Sanctions Working Group, August 1980), p. 3; and Bernard Rivers and Martin Bailey, "How Oil Seeps into South Africa," *Business and Society Review* no. 39 (Fall 1981): 54.

3. For further details, see African National Congress and South-West Africa People's Organization, "Call for an Oil Embargo" (London, March 1985), p. 6.

4. The Iranian role in South Africa is discussed in Chapters 6 and 7.

5. *Rand Daily Mail,* May 22, 1982.

6. For example, the energy supplement of *Financial Mail,* June 29, 1979, contains a wealth of information about South African oil.

CHAPTER 4

1. *Africa Contemporary Record, 1973–1974* (New York: Africana, 1974), p. B466.

2. Barbara Rogers maintains that many foreign oil companies did not anticipate significant oil finds, but participated in exploratory efforts in order to retain South Africa's good will. See *White Wealth and Black Poverty* (Westport, Conn.: Greenwood, 1976), p. 141.

3. See energy supplement of *Financial Mail*, June 29, 1979, p. 32; *Financial Mail*, January 30, 1981, p. 373; and *Windhoek Advertiser*, February 15, 1985 (*Facts and Reports* 15, no. G, March 29, 1985:5).

4. *Citizen*, January 7, 1984; and *Hansard*, May 4, 1984, p. 6.

5. *Economist* 290, no 7323 (January 7, 1984): 35.

6. *Guardian*, August 18, 1986; *Anti-Apartheid News*, April 1988 (*Facts and Reports* 18, no. H, April 8, 1988:16); and supplement to *Financial Mail*, June 20, 1986, pp. 39–40.

7. Sanctions Working Group, "Implementing an Effective Oil Embargo Against South Africa: The Current Situation" (New York: Centre Against Apartheid, August 1980), pp. 1–2; "British Petroleum's Oil Deal with South Africa" (London: Anti-Apartheid Movement, September 1980), p. 1; "South African Mining Interests Move into North Sea Oil" (London: Anti-Apartheid Movement, March 1981), pp. 1–2; *Observer*, March 15, 1981; *South*, September 1981 (*Facts and Reports* 11, no. P/Q, August 21, 1981:25); and "Oil and Apartheid: Churches' Challenge to Shell and BP" (London: Christian Concern for Southern Africa, March 1982), p. 13.

8. Willie Breytenbach, "Third World Moves Against South Africa," in Deon Geldenhuys, ed., *Sanctions Against South Africa* (South African Institute of International Affairs, December 1979), p. 26.

9. *Lloyd's List*, September 27, 1980, p. 2; *American Metal Market*, January 25, 1974; and William Robinson, "Fluor: Apartheid's Energy Partner," *Business and Society Review*, no. 37 (Spring 1981): 59–60.

10. *South Africa's Lifeline: Violations of the Oil Embargo, 1983–1984* (Amsterdam: Shipping Research Bureau, September 1986), p. 29.

11. *Financial Mail*, July 10, 1981, p. 192.

12. See Advocate-General of South Africa, "Report in Terms of Section 5(1) of the Advocate-General Act, 1979 (Act 118 of 1979)," June 27, 1984, pp. 10–13; and *South Africa's Lifeline*, pp. 35–37.

13. Energy supplement of *Financial Mail*, June 29, 1979, p. 9.

14. See Advocate-General of South Africa, "Report of Section 5(1)," p. 8.

15. See *Financial Mail*, February 21, 1986, p. 33; and *South Africa's Lifeline*, p. 39.

16. *Lloyd's List*, February 24, 1979, p. 2.

17. *Economist* 271, no. 7082 (May 26, 1979): 94.

18. *Africa Confidential* 24, no. 1 (January 5, 1983): 3.

19. Dutch Anti-Apartheid Movements, "Transnational Corporations in South Africa: The Role of Shell," PH Material no. 23 (New York: United Nations, September 13, 1985), p. 3.

20. *Financial Mail*, December 12, 1980, p. 1287; Subcommittee on African Affairs, Committee on Foreign Relations, U.S. Senate, "South Africa" (Washington: U.S. Government Printing Office, 1977), p. 390; and *South*

Africa 1974 (Johannesburg: Department of Information, 1974), p. 32.

21. *Rand Daily Mail*, October 31, 1983; Arnt Spandau, *Economic Boycott Against South Africa* (Kenwyn: Juta, 1979), p. 60; and Tony Koenderman, *Sanctions: The Threat to South Africa* (Johannesburg: Jonathan Ball, 1982), pp. 8 and 267.

22. *Newsletter on the Oil Embargo Against South Africa*, no.5 (September 1986): 3.

23. *Financial Mail*, April 11, 1980, p. 143.

24. Ibid., March 14, 1980, p. 1067.

25. *Weekly Mail* (South Africa), August 22, 1986.

26. *South Africa's Lifeline*, p. 39; earlier estimates have been $1.5 billion, $1.85 billion, and $1.99 billion. See *Lloyd's List*, July 9, 1983, p. 1; "West European Companies Breaking the Oil Embargo Against South Africa" (Amsterdam: Shipping Research Bureau, September 1985), p. 7; and African National Congress and South-West Africa People's Organization, "Call for an Oil Embargo" (London, March 1985), p. 1.

CHAPTER 5

1. Barbara Rogers, "Southern Africa and the Oil Embargo," *Africa Today* 21, no. 2 (Spring 1974): 8.

2. Oil Embargo Against South Africa," *Sechaba* 8, no. 5 (May 1974): 11.

3. See *Africa Confidential* 24, no. 7 (March 30, 1983): 1.

4. Sanctions Working Group, "Implementing an Effective Oil Embargo Against South Africa: The Current Situation" (New York: Centre Against Apartheid, August 1980), p. 75.

5. "Fuelling Apartheid: Shell and the Military" (London: Christian Concern for Southern Africa, 1984), p. 6.

6. Sanctions Working Group, "Implementing an Effective Embargo," pp. 118–19.

7. Ibid., p. 239.

8. *Financial Times*, May 21, 1985.

9. *Observer*, November 13, 1983. Tankers have occasionally sailed for Durban from Dar es Salaam, although it is unclear what role they may possibly have played in transporting oil between Tanzania and South Africa. See *Africa News*, February 16, 1983, p. 9.

10. See Chapter 11.

11. *Lloyd's List*, October 14, 1982, p. 2; January 7, 1983, p. 2; August 16, 1985, p. 3; and Sanctions Working Group, "Implementing an Effective Embargo," p. 28.

12. *The Star* (South Africa), December 2, 1986.

13. Sanctions Working Group, "Implementing an Effective Embargo," pp. 189 and 192; "Fuelling Apartheid," p. 6.

14. *Herald* (Zimbabwe), April 10, 1986 (*Facts and Reports* 16, no. K, June 6, 1986: 24).

15. *South Scan*, November 4, 1986 (ibid., no. W, November 21, 1986: 19).

16. *Times*, August 14, 1986.

17. Sanctions Working Group, "Implementing an Effective Embargo," p.

191.

CHAPTER 6

1. Martin Bailey and Bernard Rivers, "Oil Sanctions Against South Africa" (New York: Centre Against Apartheid, June 1978), p. 21; and Barbara Rogers, "Southern Africa and the Oil Embargo," *Africa Today* 21, no. 2 (Spring 1974): 3.

2. Committee on Foreign Relations, U.S. Senate, "U. S. Corporate Interests in Africa" (Washington: U.S. Government Printing Office, 1978), p. 57. South Africa was diverting some crude imports to Rhodesia during this period.

3. *Financial Mail*, November 23, 1973, p. 794.

4. Barbara Rogers, *White Wealth and Black Poverty* (Westport, Conn.: Greenwood, 1976), p. 262.

5. Committee on Foreign Relations, "U.S. Corporate Interests," p. 57.

6. Bailey and Rivers, "Oil Sanctions Against South Africa," p. 22; and *Financial Mail*, November 23, 1973, p. 794.

7. African National Congress (ANC), "Fuelling Apartheid"(New York: Centre Against Apartheid, April 1980), p. 7.

8. UN General Assembly, 18th Session, 1257th Plenary Meeting, November 13, 1963, pp. 2–3.

9. See Dirk Mostert, speech in Johannesburg, August 23, 1984, p. 4; and Advocate-General of South Africa, "Report in Terms of Section 5(1) of the Advocate-General Act, 1979 (Act 118 of 1979)," June 27, 1984, p. 7.

10. *Financial Mail*, July 19, 1974, p. 223.

11. *Financial Times*, October 30, 1978; and *Financial Mail*, November 23, 1978.

12. Michael Tanzer, Terisa Turner, Bernard Rivers, and Martin Bailey, "Oil—A Weapon Against Apartheid," International Seminar on an Oil Embargo Against South Africa, Amsterdam, March 1980, p. 31; and Terisa Turner, "Trade Union Action to Stop Oil to South Africa" (Port Harcourt, Nigeria: University of Port Harcourt, 1985), p. 9.

13. *Keesings Contemporary Archives*, July 27, 1979, p. 29740; and Martin Bailey, "The Impact on South Africa of the Cut-off of Iranian Oil" (New York: Centre Against Apartheid, July 1979), p. 1.

14. ANC, "Fuelling Apartheid," p. 7; and Martin Bailey, "Oil Sanctions: South Africa's Weak Link" (New York: Centre Against Apartheid, April 1980), p. 3.

15. For allegations, see *Lloyd's List*, January 29, 1979, p. 2. The South African government denied these assertions. See *Rand Daily Mail*, February 8, 1979.

16. *Rand Daily Mail*, January 19, 1979.

17. *Rand Daily Mail*, June 8, 1979.

18. *Financial Mail*, July 27, 1979, p. 297.

19. *Financial Times*, June 27, 1979.

20. *Keesings Contemporary Archives*, March 21, 1980, p. 30148; *Financial Mail*, September 7, 1979, p. 943; and *Financial Times*, April 20, 1979.

21. *Lloyd's List*, January 29, 1979, p. 2; Willie Breytenbach, "Third World

Moves Against South Africa," in Deon Geldenhuys, ed., *Sanctions Against South Africa* (South African Institute of International Affairs, December 1979), p. 9; and *Financial Mail,* September 17, 1979, p. 943.

22. *Business Week,* no. 2575 (March 5, 1979): 30.

23. *Newsletter on the Oil Embargo Against South Africa,* no. 8 (July 1987): 2 and *Oil to South Africa* (Amsterdam, Shipping Research Bureau, September, 1988), p. 13. See also Permanent Mission of Israel to the United Nations, "Arab Oil Trade with South Africa Exposed" (New York: November 6, 1986).

24. Letter from Iranian embassy in Oslo, Norway, August 13, 1986, in *Newsletter on the Oil Embargo Against South Africa,* no. 5 (September 1986): 12.

25. Ibid., no. 8 (July 1987): 2.

26. *Africa Confidential,* January 22, 1988; and *Observer,* October 18, 1987.

27. *International Herald Tribune,* November 27, 1987.

28. United Nations General Assembly and Security Council, "Report on the Intergovernmental Group to Monitor the Supply and Shipping of Petroleum Products to South Africa," A/42/45 S/19251 (November 5, 1987), p. 27.

CHAPTER 7

1. *Africa Confidential* 27, no. 17 (August 20, 1986): 5; ibid. 24, no. 1 (January 5, 1983): 4; and *Newsletter on the Oil Embargo Against South Africa,* no. 6 (January 1987): 14. There have been reports that Iraq and South Africa have bartered oil for arms, but they have not been adequately substantiated and may be disseminated to counter more reliable reports that Iran and South Africa engage in such transactions. See ibid., no. 4 (May 1986): 8.

2. *Oil Tankers to South Africa, 1980–81* (Amsterdam: Shipping Research Bureau, June 1982), p. 21; *Secret Oil Deliveries to South Africa, 1981–1982* (Amsterdam: Shipping Research Bureau, June 1984), p. 13; and *South Africa's Lifeline: Violations of the Oil Embargo, 1983–1984* (Amsterdam: Shipping Research Bureau, September 1986), p. 10.

3. United Nations General Assembly and Security Council, "Report on the Intergovernmental Group to Monitor the Supply and Shipping of Petroleum Products to South Africa," A/42/45 S/19251 (November 5, 1987), p. 41.

4. See *Newsletter on the Oil Embargo Against South Africa,* no. 8 (July 1987): 2; *Oil to South Africa* (Amsterdam: Shipping Research Bureau, September 1988), p. 13.

5. *Observer,* June 3, 1984.

6. Advocate-General of South Africa, "Report in Terms of Section 5(1) of the Advocate-General Act, 1979 (Act 118 of 1979)," June 27, 1984; and Robert Whitehill, "The Sanctions That Never Were: Arab and Iranian Oil Sales to South Africa," *Middle East Review* 19, no. 1 (Fall 1986): 41.

7. *The Star,* October 16, 1982; and *Sunday Times,* October 31, 1982, p. 1.

8. Subcommittee on Africa, Committee on Foreign Affairs, U.S. House of Representatives, "Possible Violation or Circumvention of the Clark Amendment," July 1, 1987, pp. 26 and 94; and *South Scan,* August 5, 1987 (*Facts and Reports* 17, no. OP, August 14, 1987: 9).

9. *South Scan,* August 5, 1987; *Newsletter on the Oil Embargo Against South*

Africa, no. 8, (July 1987), p. 1.

10. Ephraim Dowek, "Arabs Supply Oil to South Africa," International Labour Conference, Geneva, June 19, 1986 (London: Labour Friends of Israel, 1986).

11. "The Arabs Fuel South Africa" (Washington: The American Israel Public Affairs Committee, August 1984).

12. Permanent Mission of Israel to the United Nations, "Arab Oil Trade with South Africa Exposed" (New York: November 6, 1986). For further criticism of the Israeli methodology, see *Newsletter on the Oil Embargo Against South Africa*, no. 6 (January 1987): 11.

13. "How Britain Fuels the Apartheid War Machine" (London: Anti-Apartheid Movement, March 1981), p. 6.

14. "Crude Oil Deliveries to South Africa From Brunei" (Amsterdam: Shipping Research Bureau, January 1987), p. 11.

15. Bernard Rivers and Martin Bailey, "How Oil Seeps into South Africa," *Business and Society Review*, no. 39 (Fall 1981): 58.

16. "Crude Oil Delivers to South Africa From Brunei," pp. 1–2 and 6.

17. Ibid., p. 1. See also these Shirebu reports: *Oil Tankers to South Africa, 1980–81*, p. 21; *Secret Oil Deliveries to South Africa, 1981–1982*, p. 13; *South Africa's Lifeline* p. 10. And see Michael Tanzer, Terisa Turner, Bernard Rivers, and Martin Bailey, "Oil—A Weapon Against Apartheid," International Seminar on an Oil Embargo Against South Africa, Amsterdam, March 1980, p. 26; and Martin Bailey, "The Impact on South Africa of the Cut-off of Iranian Oil" (New York: Centre Against Apartheid, July 1979), p. 2.

18. *South Africa's Lifeline*, p. 19.

19. *Observer*, November 2, 1986; and "Crude Oil Deliveries to South Africa from Brunei," p. 12.

20. *Africa Now*, April 1981 (*Facts and Reports* 11, no. H, April 17, 1981: 5); *Africa Confidential* 24, no. 7 (March 30, 1983): 2; and Sanctions Working Group, "Implementing an Effective Oil Embargo Against South Africa: The Current Situation" (New York: Centre Against Apartheid, August 1980), p. 74.

21. *Africa Confidential* 24, no. 1 (January 5, 1983): 3–4; *Observer*, May 30, 1982; and *Secret Oil Deliveries to South Africa, 1981–1982*, p. 20.

22. See *Keesings Contemporary Archives*, September 22, 1978, p. 29215, and January, 1984, p. 32611.

23. *Lloyd's List, July 29, 1983, p. 1, and December 12, 1984, p. 3.*

24. *New Nigerian*, December 3, 1977 (*Africa Research Bulletin: Economic, Financial and Technical Series* 14, no. 11, December 31, 1977: 4497); and "Nigerian Oil Fraud Exposed," *New African*, no. 199 (April 1984): 11 and 14.

25. For information on the role of middlemen, see *West Africa*, March 17, 1986 (*Facts and Reports* 16, no. G, April 11, 1986: 11); *Observer*, February 4, 1979; Desaix Myers III, *U.S. Business in South Africa* (Bloomington: Indiana University Press, 1980), p. 60; Agence France-Presse, March 10, 1979; and *Natal Mercury* (South Africa), February 5, 1979. It is possible that some Nigerian oil has been routed to South Africa through Equatorial Guinea. See *New African*, January 1988 (*Facts and Reports* 18, no. C, February 12, 1988: 2).

26. See Tanzer et al., "Oil—A Weapon Against Apartheid," p. 35.

27. *The Star*, May 19, 1979; *Guardian*, May 18, 1979; *Lloyd's List*, May 18, 1979, p. 1; June 12, 1979, p. 1; and June 18, 1979, p. 1; and *Nigeria Newsletter*, May 28, 1979 (*Africa Research Bulletin: Economic, Financial and Technical Series*

16, no. 5, June 30, 1979:5125).

CHAPTER 8

1. *Economist* 276, no. 7140 (July 5, 1980): 49; Martin Bailey, "Western Europe and the South African Oil Embargo" (New York: Centre Against Apartheid, February 1981), p. 5; *Financial Times*, February 12, 1982; and Dutch General News Bureau, May 5, 1983 *(Facts and Reports* 13, no. J, May 13, 1983: 19).

2. *Oil Tankers to South Africa, 1980–81* (Amsterdam: Shipping Research Bureau, June 1982), p. 16; "West European Companies Breaking the Oil Embargo Against South Africa" (Amsterdam: Shipping Research Bureau, September 1985), p. 15; and "Oil Supplies to South Africa: The Role of Tankers Connected with the Netherlands and the Netherlands Antilles" (Amsterdam: Shipping Research Bureau, January 1981), p. 2.

3. *South Africa's Lifeline: Violations of the Oil Embargo, 1983–1984* (Amsterdam: Shipping Research Bureau, September 1986), pp. 23–26.

4. "Oil Supplies to South Africa," pp. 2, 5, and 9; "Fuelling Apartheid: Shell and the Military" (London: Christian Concern for Southern Africa, 1984), p. 16; and *Oil Tankers to South Africa 1980–1981* p. 19.

5. "Oil Shipments to South Africa by Tankers Thorsaga, Thorshavet and Thorsholm Owned by A/S Thor Dahl of Norway (1981–1984)" (Amsterdam: Shipping Research Bureau, December 1984), p. I-1; "West European Countries Breaking the Oil Embargo Against South Africa" (Amsterdam: Shipping Research Bureau, September 1985), pp. 11–12; *Secret Oil Deliveries to South Africa 1981–1982* (Amsterdam: Shipping Research Bureau, June 1984), p. 11; *Oil Tankers to South Africa, 1980–81*, p. 18; and *South Africa's Lifeline*, pp. 6–7.

6. "Oil Shipments to South Africa by Tankers Owned and Managed by Sig. Bergesen D. Y. and Co. of Norway" (Amsterdam: Shipping Research Bureau, July 1985), p. 8; and *South Africa's Lifeline*, p. 20.

7. *Secret Oil Deliveries to South Africa, 1981–1982*, pp. 24–25; Robert Whitehill, "Apartheid's Oil," *The New Republic* 194, no. 6 (February 10, 1986): 11; *Newsletter on the Oil Embargo Against South Africa* 1, no. 2 (June 1985): 4; and "John Deuss: Transworld Oil" (Amsterdam: Shipping Research Bureau, January 1985), p. 2.

8. *Times*, October 8, 1985.

9. *Lloyd's List*, July 11, 1986, p. 3.

10. *Newsletter on the Oil Embargo Against South Africa*, no. 8 (July 1987): 7.

11. Ibid., no. 7 (April 1987): 14.

12. "West European Companies," pp. 10–11.

13. *Observer* August 5, 1984; and *Newsletter on the Oil Embargo Against South Africa* 1, no. 2 (June 1985): 6.

14. *Oil Tankers to South Africa, 1980–81*, pp. 5–6; *Secret Oil Deliveries to South Africa, 1981–1982*, p. 25; "West European Involvement in Breaking the Oil Embargo Against South Africa" (Amsterdam: Shipping Research Bureau, May 1985), p. 8; *Lloyd's List*, January 20, 1981, p. 1, and January 21, 1981, p. 1; *Guardian*, January 18, 1981; *Financial Times*, January 21, 1981; and *Observer*, August 5, 1984.

15. *Lloyd's List*, January 22, 1981, p. 2.

16. *Newsletter on the Oil Embargo Against South Africa* 1, no. 2 (June 1985): 4. See also *Oil Tankers to South Africa, 1980–81*, pp. 18 and 25; *Observer*, February 22, 1981, and August 5, 1984; *Africa Confidential* 24, no. 1 (January 5, 1983): 1; and "Fuelling Apartheid," p. 16.

17. "West European Companies," p. 15.

18. "How Britain Fuels the Apartheid War Machine" (London: Anti-Apartheid Movement, March 1981), pp. 16–17. See also "West European Involvement," p. 8.

19. Martin Bailey, "Britain's Role in South Africa's Oil Supply," Information Paper no. 37, Labour Party International Department (January 1983), p. 1; and a letter from former foreign secretary David Owen to Lord Carrington in Michael Tanzer, Terisa Turner, Bernard Rivers, and Martin Bailey, "Oil—A Weapon Against Apartheid," International Seminar on an Oil Embargo Against South Africa, Amsterdam, March 1980, p. 73.

20. Stanley Uys, "Prospects for an Oil Boycott," *Africa Report* 25, no. 5 (September–October 1980): 17; "Oil and Apartheid: Churches' Challenge to Shell and BP (London: Christian Concern for Southern Africa, 1982), p. 12; and Martin Bailey, "Skilful Way with Oil Embargoes," *New Statesman* (July 6, 1979): 2.

21. *Africa Contemporary Record, 1979–1980* (New York: Africana, 1981), p. A131.

22. Tanzer et al., "Oil—A Weapon Against Apartheid," p. 73.

23. Letter from Lord Carrington, June 26, 1979, in ibid., pp. 76–77.

24. *International Herald Tribune*, August 1, 1979.

25. See Tanzer et al., "Oil—A Weapon Against Apartheid," pp. 37–38; and Terisa Turner, "Trade Union Action to Stop Oil to South Africa" (Port Harcourt, Nigeria: University of Port Harcourt, 1985), p. 11.

26. *Lloyd's List*, February 26, 1981, p. 1.

27. "Marimpex: A German Oil Supplier to South Africa" (Amsterdam: Shipping Research Bureau, October 1985), pp. 15 and 22; "West European Companies," p. 16; and *South Africa's Lifeline*, p. 14.

28. "Oil Shipments to South Africa on Maersk Tankers: The Role of A. P. Moller of Denmark" (Amsterdam: Shipping Research Bureau, June 1985), pp. 1, 4, and 16; and "West European Companies," p. 13. For Danish statistics on crude delivered to South Africa on Danish tankers, see *Newsletter on the Oil Embargo Against South Africa*, 1, no. 2 (June 1985): 6.

29. *Newsletter* 1, no. 2 (June 1985): 2; *Observer*, May 19, 1985; and "West European Companies," p. 10.

30. *Lloyd's List*, May 17, 1984, p. 3, and September 17, 1985, p. 1.

31. *South Africa's Lifeline*, p. 48; *New York Times*, December 8, 1985, p. 9; and *Lloyd's List*, December 7, 1985, p. 6; December 9, 1985, p. 1; and December 10, 1985, p. 8.

32. Sanctions Working Group, "Tracking Oil to South Africa," *Arab Oil and Gas* (December 1, 1980): 19; *Oil Tankers to South Africa, 1980–81*, p. 18; and *South Africa's Lifeline*, pp. 64–67.

33. Whitehill, "Apartheid's Oil," p. 11; Advocate-General of South Africa, "Report in Terms of Section 5(1) of the Advocate-General Act, 1979 (Act 118 of 1979), June 27, 1984, p. 16; "John Deuss: Transworld Oil" (Amsterdam: Shipping Research Bureau, January 1985), p. 1B; *Newsletter on the Oil Embargo Against South Africa* 1, no. 2, (June 1985): 1; "West European

Companies," p. 14; and *South Africa's Lifeline*, p. 12.

34. *Observer*, August 5, 1984; and Advocate General of South Africa, "Report of Section 5(1)," p. 42.

35. *De Volkskrant*, (Netherlands) January 8, 1985.

36. *Observer*, October 26, 1986.

37. *International Herald Tribune*, October 15, 1987.

38. *Observer*, August 5, 1984; "West European Companies," p. 15; Advocate-General of South Africa, "Report of Section 5(1)," p. 39; and "Crude Oil Deliveries to South Africa from Brunei" (Amsterdam: Shipping Research Bureau, January 1987), p. 6.

39. "Deliveries From Brunei," p. 12.

40. *New York Times*, January 28, 1984, p. 40; and *Rand Daily Mail*, March 21, 1985.

41. *Hansard*, April 26, 1984, p. 5218, and April 27, 1984, pp. 5294–97; *Newsletter on the Oil Embargo Against South Africa* 1, no. 1 (February 1985): 3; *Secret Oil Deliveries to South Africa, 1981–1982*, p. 4; *Financial Mail*, May 11, 1984, p. 30; *Times*, April 30, 1984; and *Rand Daily Mail*, July 13, 1984.

42. *The Star*, June 25, 1984; *Newsletter on the Oil Embargo Against South Africa* 1, no. 1 (February 1985): 3; and Advocate-General of South Africa, "Report of Section 5(1)," p. 42.

43. *Observer*, June 3, 1984.

44. *Sunday Times*, April 29, 1984; *Guardian*, April 30, 1984; and *Hansard*, May 3, 1984, pp. 5693–94. According to the *Rand Daily Mail* of September 12, 1983, Chiavelli may have contributed money for the ruling National Party's campaign in 1983 regarding a referendum on constitutional changes.

45. *Hansard*, May 4, 1986, p. 86; and *Rand Daily Mail*, April 30, 1984.

46. *Rand Daily Mail*, February 11, 1985.

47. *Observer*, August 9, 1981.

48. *Rand Daily Mail*, August 19, 1982.

49. Advocate-General of South Africa, "Report of Section 5(1)," p. 40.

50. *Observer* August 9, 1981; and *Sunday Express*, May 6, 1984.

51. *Africa Confidential* 24, no. 1 (January 5, 1983): 2; and *Observer*, August 9, 1981.

52. Advocate-General of South Africa, "Report of Section 5(1)," and *Sunday Tribune*. June 10, 1984. See also *Hansard*, April 25, 1984, p. 5077.

53. Advocate-General of South Africa, "Report of Section 5(1)," p. 36; *Financial Mail*, March 2, 1984, p. 45: *Rand Daily Mail*, March 6, 1984; *Guardian*, April 30, 1984; *Daily News*, April 26, 1984; and Robert Whitehill, "The Sanctions That Never Were: Arab and Iranian Oil Sales to South Africa," *Middle East Review* 19, no. 1 (Fall 1986): 42–43.

54. *Hansard*, May 20, 1985, p. 5856.

CHAPTER 9

1. "Oil Shipments to South Africa on Maersk Tankers: The Role of A. P. Moller of Denmark" (Amsterdam: Shipping Research Bureau, June 1985), p. 10.

2. "Marimpex: A German Oil Supplier to South Africa" (Amsterdam:

Shipping Research Bureau, October 1985), p. 9; *Africa Business*, March 1982; "John Deuss: Transworld Oil" (Amsterdam: Shipping Research Bureau, January 1985), p. 4B; *Times*, September 12, 1986, p. 10; and *Oil Tankers to South Africa, 1980–81* (Amsterdam: Shipping Research Bureau, June 1982), p. 32.

3. "John Deuss," pp. 3B and 3C; *Africa News*, February 16, 1984, p. 9; and Arthur Jay Klinghoffer, *Fraud of the Century: The Case of the Mysterious Supertanker Salem* (London: Routledge, 1988), pp. 33–34.

4. Klinghoffer, *Fraud of the Century*, p. 29; *Guardian*, August 6, 1980; "John Deuss," p. 2; *Dagbladet*, (Netherlands) April 12, 1980; and *Verdens Gang* (Netherlands), May 24, 1980.

5 *Times*, September 12, 1986, p. 10.

6. "Oil Shipments to South Africa on Maersk Tankers," p. 3.

7. *Newsletter on the Oil Embargo Against South Africa*, no. 5 (September 1986): 2.

8. "Oil Shipments to South Africa on Maersk Tankers," pp. 14–15 and Annex A; "John Deuss," p. 4B; and *Times*, September 12, 1986.

9. *Secret Oil Deliveries to South Africa, 1981–1982* (Amsterdam: Shipping Research Bureau, June 1984), p. 20; and Sanctions Working Group, "Implementing an Effective Oil Embargo Against South Africa: The Current Situation," (New York: Centre Against Apartheid, August 1980), p. 3.

10. Sanctions Working Group, "Implementing an Effective Oil Embargo," p. 4.

11. *Lloyd's List*, April 23, 1981, p. 10; April 27, 1981, p. 16; April 28, 1981, p. 16; and June 17, 1981, p. 1.

12 Oliver Tambo, Address to Conference of Maritime Unions Against Apartheid, London, October 30, 1985, p. 4.

13. See "Maritime Fraud," *Lloyd's Shipping Economist* 8, no. 2 (February 1986): 8.

14. "SFF Association Memorandum on the Salem Tanker at Request of Minister of Mineral and Energy Affairs," March 9, 1983, p. 8.

15. See Klinghoffer, *Fraud of the Century*, p. 73.

16. *South Africa's Lifeline: Violations of the Oil Embargo, 1983–1984* (Amsterdam: Shipping Research Bureau, September 1986), pp. 17, 62–63, and 75; and "Crude Oil Deliveries to South Africa from Brunei" (Amsterdam: Shipping Research Bureau, January 1987), p. 20. In August 1985 Sanko filed for bankruptcy and sold most of its tankers.

17. "Oil Shipments to South Africa by the Tankers Thorsaga, Thorshavet and Thorsholm Owned by A/S Thor Dahl of Norway (1981–1984)" (Amsterdam: Shipping Research Bureau, December 1984), Annex D.

18. *Oil Tankers to South Africa, 1980–81*, p. 12.

19 Tony Koenderman, *Sanctions: The Threat to South Africa* (Johannesburg: Jonathan Ball, 1982), p. 267.

20. *Observer*, January 18, 1981.

21. "Oil Shipments to South Africa on Maersk Tankers," p. 8.

22. Klinghoffer, *Fraud of the Century*, pp. 76–77.

23. *Lloyd's List*, April 7, 1980, p. 1, and October 14, 1980, p. 1; "Oil Sanctions Against South Africa" (New York: Sanctions Working Group, August 1980), p. 8; Bernard Rivers and Martin Bailey, "How Oil Seeps into South Africa," *Business and Society Review*, no. 39 (Fall 1981): 58; and Bureau of Maritime Affairs, Republic of Liberia, "Report on the Formal Investigation into the Albahaa B." Monrovia, August 5, 1982.

24. "West European Companies Breaking the Oil Embargo Against South Africa" (Amsterdam: Shipping Research Bureau, September 1985), p. 12; and "Oil Shipments to South Africa on Maersk Tankers," p. 14.

25. "Maersk Tankers," p. 13; and *Secret Oil Deliveries to South Africa, 1981–1982*, p. 35

26. *Secret Oil Deliveries*, p. 19; Bureau of Maritime Affairs, Republic of Liberia, "Report of the Marine Board of Investigation into the Collision of the Liberian Flag VLCC Aegean Captain and the Greek Flag VLCC Atlantic Empress," July 24, 1981, pp. i and 15; and *Africa Confidential* 24, no. 1 (January 5, 1983): 2.

27. "Report on the Aegean Captain and Atlantic Empress," pp. ii, iii, and v; and *Lloyd's List*, March 15, 1986, p. 1.

28. *De Volkskrant*, September 15, 1981; and George Grayson, "The San Jose Oil Facility: South-South Cooperation," *Third World Quarterly* 7, no. 2 (April 1985): 398.

29. *Independent*, October 23, 1987, and November 11, 1987.

30. *Africa Confidential* 28, no. 8 (April 15, 1987): 1; and ibid. 28, no. 12 (June 10, 1987): 8.

31. David Taylor, "South African Connection," *The Listener* (May 24, 1979); BBC 1, "Panorama," May 21, 1979, p. 6; and "Oily Intrigue ... Bungled Invasion," *New African*, no. 143 (July 1979): 11.

32. "Oily Intrigue" and "Panorama," p. 8.

33. Taylor, "South African Connection,"; "Panorama," pp. 9–11; and "Oily Intrigue," p. 11.

34. "Oily Intrigue," pp. 10–11; and "Panorama," p. 12.

35. *Keesings Contemporary Archives*, May 30, 1980, p. 30269; and "Oily Intrigue" p. 13

36. Michael Tanzer, Terisa Turner, Bernard Rivers, and Martin Bailey, "Oil—A Weapon Against Apartheid," International Seminar on an Oil Embargo Against South Africa, Amsterdam, March 1980, p. 40; and Taylor, "South African Connection."

37. "South African Connection"; "Panorama," p. 16; and "Oily Intrigue," p. 13.

38. "Oily Intrigue," pp. 13–14; and "Panorama," pp. 15–16.

39. "Panorama" p. 16; "Oily Intrigue," p. 13; Michael Tanzer, Terisa Turner, Jennifer Davis, and Sybil Wong, "Toward an Effective Oil Embargo of South Africa," *Monthly Review* 32, no. 7 (1980): 59; and *Africa Contemporary Record, 1979–80* (New York: Africana, 1981), p. B868.

40. "Oily Intrigue," p. 13.

41. Tanzer et al., "Oil—A Weapon Against Apartheid," p. 40.

42. *Keesings Contemporary Archives*, May 30, 1980, p. 30269.

43. "Panorama," pp. 14 and 17; and Tanzer et al., "Oil—A Weapon Against Apartheid," p. 40.

44. *Keesings Contemporary Archives*, May 30, 1980, p. 30269 and Tanzer et al., "Toward an Effective Oil Embargo of South Africa," p. 59.

CHAPTER 10

1. For a thorough examination of the *Salem* case, see Arthur Jay

Klinghoffer, *Fraud of the Century: The Case of the Mysterious Supertanker* Salem (London: Routledge, 1988).

2. Calculations pertinent to the fraud may be found in ibid. pp. 123–124.

3. P. C. Swanepoel, "Report on the Circumstances Surrounding the Sale of the Cargo of Crude Oil on Board the Vessel SALEM to South Africa in 1979," November 18, 1983, pp. 37–39 and 43; and U.S. government's trial memorandum, U.S. District Court for the Southern District of Texas, Houston Division, *U.S. v. Soudan,* p. 6.

4. Trial Memorandum, p. 8; and minutes of October 15 meeting at Sasolburg, Exhibit 864B, *U.S. v. Soudan.*

5. *U.S. v. Soudan* Exhibit 864B; and Swanepoel, "Reports," pp. 43-50.

6. *Lloyd's List,* February 6, 1986, p. 1, and February 7, 1986, pp. 1 and 10; *Deegblad Scheepvaart* (Netherlands), February 6 and 7, 1986; interview with Peter Griggs, December 15, 1986; *U.S. v. Soudan,* vol. 1, p. 25; and Swanepoel, "Reports" pp. 102, 107, and 276.

7. *U.S. v. Soudan,* vol. 1, p. 30; Griggs interview; Griggs report of June 10, 1980, as presented at the trial of Soudan, p. 41; Barbara Conway, *The Piracy Business* (Middlesex: Hamlyn, 1981), p. 74; Swanepoel, "Reports," p. 130; and statement by Tunisian seaman, January 1980, p. 5.

8. "SFF Association Memorandum on the Salem Tanker at Request of Minister of Mineral and Energy Affairs," March 9, 1983, p. 14.

9. *U.S. v. Soudan,* vol. 22, p. 172; and speech by Dirk Mostert, Johannesburg, August 23, 1984, p. 12.

10. Swanepoel, "Reports."

11. Interviews with Peter Griggs, December 15, 1986, and March 2, 1987; and M. Seward, "Salem Scam," *Oceans* (November 1985): 19.

12. *Hansard,* March 9, 1983, pp. 2630–31 and Barbara Conway, *The Piracy Business* (Middlesex: Hamlyn, 1981).

13. *Hansard,* April 27, 1984, p. 5295.

14. Advocate-General of South Africa, "Report in Terms of Section 5(1) of the Advocate-General Act, 1979 (Act 118 of 1979), June 27, 1984, p. 43.

15. *Hansard,* May 20, 1985, pp. 5855–57.

CHAPTER 11

1. Leonard Kapungu, *The United Nations and Economic Sanctions Against Rhodesia* (Lexington, Mass.; Lexington Books, 1973), p. 8.

2. Martin Bailey, "Ian Smith Will Testify Against Shell and BP," *New Statesman* 101, no. 2603 (February 6, 1981): 3; Jorge Jardim, *Sanctions Double-Cross: Oil to Rhodesia* (Bulawayo: Books of Rhodesia, 1979), pp. 17 and 21; and Martin Bailey, *Oilgate: The Sanctions Scandal* (London: Coronet, 1979), pp. 111–12 and 115.

3. Bailey, *Oilgate,* p. 120; and Martin Bailey and Bernard Rivers, "Oil Sanctions Against South Africa" (New York: Centre Against Apartheid, June 1978), p. 79.

4. Kapungu, *The United Nations,* p. 8.

5. Bailey, *Oilgate,* pp. 122–123 and 125.

6. Brian White, "Britain and the Implementation of Oil Sanctions Against Rhodesia," in Steve Smith and Michael Clarke, eds., *Foreign Policy*

Implementation (London: Allen and Unwin, 1985), pp. 35–37.

7. Martin Bailey, "I Never Tried to Oust Smith," *New Statesman* 101, no. 2619 (May 29, 1981): 3.

8. David Owen, *The Politics of Defence* (London: Jonathan Cape, 1972), pp. 19 and 119–23.

9. *Oilgate*, p. 157.

10. *Daily Telegraph*, December 13 and 14, 1978.

11. See Donald Losman, *International Economic Sanctions* (Albuquerque: University of New Mexico Press, 1979).

12. Harry Strack, *Sanctions: The Case of Rhodesia* (Syracuse, N.Y.: Syracuse University Press, 1978), pp. 132 and 134; and Bailey, *Oilgate*, p. 144.

13. White, "Britain and Implementation," p. 44.

14. *Financial Mail*, January 18, 1974, p. 156; Strack, *Sanctions*, p. 44; and Bailey, *Oilgate*, p. 143. Shortly after UDI, oil companies may have shifted supplies from Zambia to Rhodesia to help blunt sanctions, which must have compounded Zambia's fuel emergency. See *Africa*, no. 112 (December 1980): 19.

15. Jardim, *Double-Cross*, p. 15.

16. Martin Bailey, "Shell and BP in South Africa" (London: Anti-Apartheid Movement and Haslemere Group, 1977), p. 31; and Bailey, *Oilgate*, p. 220.

17. Bailey, *Oilgate* pp. 134–35.

18. "Oil Sanctions and Southern Africa" (London: Commonwealth Secretariat, April 1974), p. 21; and Bailey, *Oilgate*, pp. 159 and 266.

19. Bailey, *Oilgate*, p. 266; Stan Luxenberg, "American Oilgate," *The New Republic* 180, no. 5 (February 3, 1979): 20; UN Centre on Transnational Corporations, "The Activities of Transnational Corporations in the Industrial, Mining and Military Sectors of Southern Africa" (New York: United Nations, 1980), p. 32; Martin Bailey, "Now What, Dr. Owen?" *New Statesman* 97, no. 2503 (March 16, 1979): 348; *Financial Mail*, September 15, 1978, p. 956; and *Keesings Contemporary Archives*, February 9, 1979, p. 29440.

20. *Times*, October 20, 1977.

21. *Financial Mail*, January 12, 1979, and February 9, 1979; *Africa* (May 1979): 43 and *NRC-Handelsblad* (Netherlands), January 10, 1979 (*Facts and Reports* 9, no. B, January 19, 1979:16). In the Netherlands, the Scholten Commission investigated Shell's role in supplying Rhodesia through Mozambique.

22. See Martin Bailey, "Above the Law," *New Statesman* 99, no. 2565 (May 23, 1980): 771–772.

23. See Tony Koenderman, *Sanctions: The Threat to South Africa* (Johannesburg: Jonathan Ball, 1982), pp. 111 and 115.

24. See "South Africa: The Case for Mandatory Economic Sanctions," World Conference for Action Against Racist South Africa, Paris, June 1986, p. 9.

25. Ibid.

CHAPTER 12

1. Muriel Grieve, "Economic Sanctions: Theory and Practice,"

International Relations (October 1968): 443.

2. See Otto Wolf von Amerongen, "Economic Sanctions as a Foreign Policy Tool?" *International Security* 5, no. 2 (Fall 1980): 165.

3. See Margaret Doxey, "Oil and Food as International Sanctions," *International Journal* 36, no. 2 (Spring 1981): 325.

4. See Margaret Doxey, *Economic Sanctions and International Enforcement* (London: Oxford University Press, 1971), pp. 139–40.

5. Gary Hufbauer and Jeffrey Schott, *Economic Sanctions in Support of Foreign Policy Goals* (Washington: Institute for International Economics, October 1983), p. 79; and Shaheen Ayubi, Richard Bissell, Nana Amu-Brafih Korsah, and Laurie Lerner, *Economic Sanctions in U.S. Foreign Policy* (Philadelphia: Foreign Policy Research Institute, 1982), p. 81.

6. See Grieve, "Theory and Practice," p. 441.

7. *Financial Mail* energy supplement, June 29, 1979, p. 9.

8. See United Nations General Assembly and Security Council, "Report on the Intergovernmental Group to Monitor the Supply and Shipping of Petroleum Products to South Africa," A/42/45 S/19251 (November 5, 1987).

9. See Margaret Doxey, "International Sanctions: A Framework for Analysis with Special Reference to the UN and Southern Africa," *International Organization* 26, no. 3 (Summer 1972): 543–44.

10. "Oil and Food," p. 333.

11. See Brian Lapping, "Oil Sanctions Against South Africa," in Ronald Segal, ed., *Sanctions Against South Africa* (Baltimore: Penguin, 1964), p. 150.

12. Anna Schreiber, "Economic Coercion as an Instrument of Foreign Policy," *World Politics* 25, no. 3 (April, 1973): 413.

— BIBLIOGRAPHIC NOTE —

No books, and few articles, have been written on oil sanctions against South Africa. Monographs and journal articles on broader aspects of economic warfare against the Pretoria regime are therefore useful in constructing an overall conceptual context, but the details have to be garnered from more specialized materials. The best sources of information are the publications of the Shipping Research Bureau in Amsterdam, which include vital statistical data and charts chronicling the movement of tankers. Also illuminating are reports issued by the UN Centre Against Apartheid in New York and the Anti-Apartheid Movement in London, as well as papers presented at international antiapartheid conferences in Paris (May 1981), Amsterdam (September 1985), and Oslo (June 1986). External journalistic coverage is most complete in England, with *Lloyd's List* being the premier publication in this regard.

Positions of the African National Congress may be gleaned form that organization's journal in exile, *Sechaba*. South African sources are constrained by oil secrecy legislation, but the *Financial Mail* is still eminently useful, as is the *Hansard*, which presents the transcripts of parliamentary sessions. South African newspapers provide significant insights; the simplest way to locate pertinent articles is to read *Facts And Reports*, a clipping compilation of the Holland Committee on Southern Africa. Nonpublic materials are also essential, such as the report of South Africa's advocate-general on oil imports, and the P.C. Swanepoel report, which delves into the case of the supertanker *Salem*.

INDEX

111